RENEWALS 458-4574.

DATE DUE

WN
RIES

GAYLORD   PRINTED IN U.S.A.

Thomas Huxley

Dubbed "Darwin's Bulldog" for his combative role in the Victorian controversies over evolutionary theory, Thomas Huxley has been widely
regarded as the epitome of the professional scientist who emerged in
the nineteenth century from the restrictions of ecclesiastical authority
and aristocratic patronage. Yet from the 1850s until his death in 1895,
Huxley always defined himself as a "man of science," a moral and religious figure, not a scientist. Exploring Huxley's relationships with his
wife, fellow naturalists, clergymen, and men of letters, White presents a
new analysis of the authority of science, literature, and religion during
the Victorian period, showing how these different practices were woven
into a fabric of high culture and integrated into institutions of print,
education, and research. He provides a substantially different view of
Huxley's role in the evolution debates and of his relations with his scientific contemporaries, especially Richard Owen and Charles Darwin.

Paul White is an affiliated research scholar in the Department of History
and Philosophy of Science at the University of Cambridge and an editor
of *The Correspondence of Charles Darwin*.

Cambridge Science Biographies

André-Marie Ampère, Enlightenment and Electrodynamics
   JAMES R. HOFMANN
Charles Darwin, The Man and His Influence PETER J. BOWLER
Humphry Davy, Science and Power DAVID KNIGHT
Galileo, Decisive Innovator MICHAEL SHARRATT
Antoine Lavoisier, Science, Administration and Revolution
   ARTHUR DONOVAN
Henry More, and the Scientific Revolution A. RUPERT HALL
Isaac Newton, Adventurer in Thought A. RUPERT HALL
Mary Somerville, Science, Illumination, and the Female Mind
   KATHRYN A. NEELEY
Justus von Liebig, The Chemical Gatekeeper WILLIAM H. BROCK

THOMAS HUXLEY

MAKING THE "MAN OF SCIENCE"

PAUL WHITE

CAMBRIDGE
UNIVERSITY PRESS

PUBLISHED BY THE PRESS SYNDICATE OF THE UNIVERSITY OF CAMBRIDGE
The Pitt Building, Trumpington Street, Cambridge, United Kingdom

CAMBRIDGE UNIVERSITY PRESS
The Edinburgh Building, Cambridge CB2 2RU, UK
40 West 20th Street, New York, NY 10011-4211, USA
477 Williamstown Road, Port Melbourne, VIC 3207, Australia
Ruiz de Alarcón 13, 28014 Madrid, Spain
Dock House, The Waterfront, Cape Town 8001, South Africa

http://www.cambridge.org

First published 2003

Printed in the United Kingdom at the University Press, Cambridge

Typeface Palatino 10/12 pt.　　*System* LATEX 2_ε　[TB]

A catalog record for this book is available from the British Library.

Library of Congress Cataloging in Publication Data

White, Paul, 1961–
Thomas Huxley : making the "man of science" / Paul White.
　　p.　cm. – (Cambridge science biographies)
Includes bibliographical references and index.
ISBN 0-521-64019-9 – ISBN 0-521-64967-6 (pbk.)
1. Huxley, Thomas Henry, 1825–1895.　2. Evolution (Biology)–England–History–19th
century.　3. Religion and science–England–History–19th century.　4. Literature and
science–England–History–19th century.　5. Naturalists–England–Biography.　I. Title.
II. Series
QH31 .H9 W55　2003
570′.92–dc21
[B]　　　　　　　　　　　　　　　　　　　　　　　　　　　　　2002067654

ISBN 0 521 64019 9 hardback
ISBN 0 521 64967 6 paperback

For Emma

Contents

Illustrations

Cover: Thomas Huxley, c. 1861.

Acknowledgments

Access to and assistance in locating primary materials was generously extended by librarians and archivists at the following institutions: The American Philosophical Society, The Huntington Library, Senate House and the Department of Education Libraries (University of London), Greater London Public Records Office, The Wellcome Library, Cambridge University Library, and Imperial College of Science, Technology, and Medicine. A special thanks to Anne Barrett, archivist at Imperial College, for her help with the Huxley papers, and to Sir Andrew Huxley and Virginia H. Huxley for their permission to use the lovely portrait of Henrietta Heathorn.

The rich interdisciplinary culture at the University of Chicago was important in laying the foundation (though not in setting the limits!) of the project. Workshops in the history of the human sciences and in European studies were lively forums for the presentation of drafts of several chapters. Staff and students in the Morris J. Fishbein Center for the History of Science and Medicine provided a welcoming environment at the beginning of the enterprise. Robert Richards supervised the work from its inception as a Ph.D. thesis and showed great patience over the course of its many permutations and sometimes protacted periods of "gestation." Lily Kay gave much needed encouragement and criticism during the early stages of the project. Lorraine Daston read the work at various junctures, served as a thesis examiner, and helped in framing the problems of scientific persona and identity. Tom Heyck at Northwestern University shared his interest in Huxley and knowledge of Victorian intellectual debates. I am also grateful for conversations with Douglas Alchin, James Chandler, Jan Goldstein, Ron Inden, Sherry Lyons, Kavita Philip, and George Stocking.

For helping to bring a renewed enthusiasm and focus to the work, I am indebted to the community of scholars connected with the Department of History and Philosophy of Science at the University of Cambridge. Spirited criticism came from the Cabinet of Natural History and the Historiography Groups. Mario Di Gregorio, Jim Moore, Simon Schaffer, and Alison Winter commented on portions of the manuscript. Nick Jardine kindly read the whole and provided helpful suggestions. Many hours of pleasurable company and discussion with Anne and Jim Secord have been an invaluable support, and my heavy debts to their respective work are evident on the pages that follow.

The last revisions to the manuscript were completed with the aid of a fellowship at the Max-Planck-Institut für Wissenschaftsgeschichte, Berlin, and I owe special thanks to my colleagues on the Darwin Correspondence Project for granting me a leave of absence at a difficult stage. In preparing the manuscript for publication, I am grateful to Alex Holzman and Mary Child at Cambridge University Press, to Bernie Lightman and David Knight for their useful comments on revision, and to Adrian Desmond for his generous criticism and the vigorous example of his own work.

Without the friendship and encouragement of Dan Beaver and Cheryce Kramer over many years, the research and writing would have been far more onerous. For my father, who only rarely questioned the value of such scholarly endeavor, I can but offer this book, such as it is, to his memory. For my mother, who has been a constant source of strength, I know that its long anticipated completion comes as a great relief. Over countless hours, Emma Spary brought form and sense to tangled thoughts that often refused to flow from my head. Her efforts have been compensated, at least, with a few good laughs at some of my overwrought prose, and with the happy prospect of a life together – without Huxley.

Thomas Huxley

Introduction

In 1894, Thomas Huxley wrote to the editor of *Science-Gossip* magazine, criticizing the appearance in its pages of a vulgar Americanism – the word "scientist."[1] For Huxley, the term denoted the sort of technical practitioner who was valued in a nation ruled solely by concerns for utility. Such a nation, he suggested, was so culturally impoverished that it fabricated words like "electrocution" (coined from "electricity" and "execution"), thereby associating science with an instrument of death, simply for linguistic economy. "Scientist," he implied, was undignified for a person of his caliber, and improper for the community of which he was a member – men of broad learning and moral gravity, capable of pronouncing on matters of general interest. From the mid-1840s, the expression that he and other professional practitioners had used for self-designation was "man of science." It was a title that, in common with those denoting other cultural leaders of the period, such as men of letters or clergymen, was free from the connotations of intellectual or commercial narrowness that could prevent men in Victorian England from entering elite circles of learning. As a community, Victorian men of science may have differed from the "natural philosophers" of the eighteenth and early nineteenth centuries in their sharper sense of distinction from other forms of learned activity (such as literature) and in

---

[1] *Science-Gossip* 1 (1894): 243. Other objections to the term were raised by John Lubbock, Alfred Russel Wallace, Lord Rayleigh, and the duke of Argyll. The term was in fact of English origin. For a discussion of nineteenth-century usages of "science" and "scientist," see Ross 1962.

their antipathy toward patronage. But there were also important continuities with this older persona, namely a dedication to liberal education, to moral and religious foundations, and to a broad public mission.[2]

A leading expert in marine zoology in the 1850s, Huxley became a notorious figure in the debates over evolutionary theory that arose after the publication in 1859 of Charles Darwin's *Origin of Species*. In the course of his teaching at the School of Mines in London and his ardent defenses of Darwin, Huxley took up paleontology, primate anatomy, and physical anthropology; wrote the controversial book *Man's Place in Nature;* and engaged in a series of disputes with the comparative zoologist Richard Owen, one of the most eminent scientific critics of Darwin's theory of natural selection. Huxley was also extremely active in educational reform, helping to install the laboratory as essential for science teaching and lobbying successfully for the incorporation of scientific subjects into English schools and universities. Through his writings on religion, politics, and culture, he tried to extend the role of the sciences in other spheres as well, and by the end of his career Huxley had become an acknowledged public authority on matters as various as natural rights, the history of Christianity, and the relations of capital and labor.

In each of these areas, Huxley placed himself at the forefront of debates in which the meaning of science was shaped before various audiences. Moreover, the extreme difficulty he had in establishing his career, the extensive controversies in which he engaged, and the ambitious philosophic role that he adopted made explicit many of the assumptions and concerns of scientific practitioners in the period. As he came to occupy important positions within both the scientific community and the government, he was able to exercise considerable influence on the vocational goals and choices of others. Popular and widely acknowledged not only as a leader of the scientific community but also as a man of letters and an educator, Huxley came to embody the ideal of the "man of science" for a wide range of publics. Utilizing the extensive record that he left of his experiences as a science practitioner, popularizer, and debater, and of his reflections on his vocation and its social significance, this book examines the creation of the Victorian "man of science" – a persona about which surprisingly little is known. Rather than a straightforward recounting of Huxley's career, this book explores his wide-ranging activities in shaping the scientific practitioner as a historical subject. By focusing on the making of scientific identity, a number of the prevailing

---

[2] On the gentlemanly codes and broad civic concerns of scientific practitioners in the first half-century, see Kargon 1977, Berman 1978, Cannon 1978, Morrell and Thackray 1981, and Alborn 1996.

interpretations of Huxley's life may be incorporated, even while some of their underlying assumptions are recast.[3]

Huxley is among those figures most widely referred to in histories of Victorian science and, indeed, in more general studies of Victorian culture.[4] Assessments of his work have been quite varied. He has been hailed as a leading promoter of meritocracy, a tireless opponent of clerical and aristocratic forms of authority, and a progressive representative of workers and women.[5] He has also been portrayed as the architect of a new elitism of experts, the high priest of a religion of science, and an ideologue of middle-class patriarchy and European racial superiority.[6] From the 1970s, scholars have interpreted Huxley's wide-ranging career most often through the category of professionalization. Their accounts have situated Huxley among the leaders of a rising scientific community struggling against an older, priestly caste for cultural hegemony.[7] Thus, for example, the activities of Huxley's "X Club" – a small group of practitioners that began to meet regularly in London in the 1860s – have been viewed as a strategic campaign to wrest control of the scientific world from clerical and theological dominion.[8] This professionalization model has been taken up anew, although in considerably modified form, in a recent book by Adrian Desmond. The first large-scale biography of Huxley since the respectful *Life and Letters* written by Huxley's son Leonard in 1900, Desmond's lively work locates the professional struggles of scientific practitioners within a broader framework of conflict between the rising industrial Nonconformist classes and the Anglican gentry and aristocracy.[9]

Together with Desmond's biography and much recent work on Huxley, this book rests upon a large body of literature in the social

---

[3] Compare, for example, the account of courtly life as "self-fashioning" in Greenblatt 1980, and the model of the scientific self as composed of different "cultural resources" in Shapin 1991. For a more extended discussion of the approach taken in this book, see the Conclusion.

[4] Among histories of Victorian science in which Huxley figures prominently, see, for example, Allen 1978, Knight 1996, and Lightman ed. 1997; among intellectual and cultural histories, see Houghton 1957, Brantlinger ed. 1989, and Hoppen 1998. For an overview of the specialist literature on Huxley, see White 2000.

[5] Huxley's anticlericalism has been emphasized, for example, by L. Huxley ed. 1900, and Irvine 1959, and his social progressivism by Bibby 1959, Paradis 1978, and Jensen 1991.

[6] On the religion of science, see Lightman 1987; on Huxley as a defender of Victorian patriarchy, see E. Richards 1983 and 1997; on Huxley's racial theory, see Di Gregorio 1984, and Brantlinger 1997.

[7] See especially the articles collected in Turner 1993.

[8] On the "X Club," see R. MacLeod 1970b, and Barton 1990 and 1998.

[9] See especially Desmond 1998: 615–43, for an overview of the author's approach. See also Desmond 2001.

history of Victorian science. Based on studies of the membership of learned societies, publication in specialist journals, paid positions in teaching and research, and correspondence networks, previous studies have provided a fair picture of a coherent scientific community that began to form in the middle decades of the nineteenth century.[10] This book is also particularly indebted to more general studies of the Victorian period, which have sought to locate the emerging scientific community in a larger constellation of learned groups and intellectual disciplines.[11] Much of this literature has assumed, however, that the identity being shaped through professionalization was that of the "scientist," that is, a highly trained expert who derived an income from research. Such studies have often argued for the increasing autonomy of scientific practice in the Victorian period, alongside the steady advance of scientific authority within other social and cultural domains.[12] Huxley frequently features in these interpretations as the epitome of the rising, professional scientist. Yet, according to his own account, he was neither a scientist nor a professional in the modern sense; nor was the salaried expert the preeminent authority figure in the period, even on matters scientific.[13]

When viewed as a problem of identity formation rather than of professionalization, Huxley's career and the categories that have been used to interpret it appear in a different light. Indeed, the boundaries of scientific identity remained permeable right through the Victorian period – and it is this permeability, rather than professional autonomy, that was crucial to Huxley's authority as a "man of science." If Huxley was successful in advancing the place of science in Victorian society, this was largely a result of his ability to maintain links and forge new alliances with groups other than specialist researchers. These alliances were not merely strategic but were woven into the very fabric of scientific identity. Rather than embrace a narrowly bounded definition of science such as might guarantee exclusiveness and autonomy, Huxley worked hard to bind the meaning of science to values and practices derived from other cultural domains. For example, he made science reliant on moral codes drawn from domesticity and gentlemanliness; he allied science

[10] Among the earlier social studies of the Victorian scientific community, see Mendelsohn 1964, Roderick 1967, Morrell 1971, Cardwell 1972, and Thackray 1974. See also n. 2 above.

[11] Raymond Williams 1958, Burrow 1966, Reader 1966, Cannon 1978, Heyck 1982, R. MacLeod ed. 1988, Perkin 1989.

[12] See, for example, the revealingly titled *All Scientists Now* (Hall 1984).

[13] For arguments against the application of twentieth-century, sociological models of professionalization to the nineteenth century, see Geison 1978, Alter 1987, and Goldstein 1987.

with extant cultural spheres, including literature and religion; and he appealed to models of labor and progress familiar to artisanal and industrial audiences. Such strategies tied science to cultural forms that carried great weight during the Victorian period. This process of definition was mutual; that is, it involved extensive networks and engagements with men of letters, clergymen, workingmen, and -women who, in fashioning themselves, helped to shape the meaning of science. This book is thus organized around the cultural practices and social groups that figured most prominently in Huxley's scientific life. But it also examines the active processes of categorization that went on in defining these practices and groups, both in public debates and in a whole range of private relations and activities. Problems of cultural identity and authority were negotiated not only in print or in grand auditoriums, such as the famous Oxford meeting of the British Association in 1860 where Huxley squared off against the bishop of Oxford, but also in everyday life and in domestic relations, friendships, and correspondence.

The term "scientist" was not of American origin as Huxley had supposed. It had been coined by the Cambridge don William Whewell in the 1830s in order to consolidate what he considered to be an increasingly heterogeneous and fragmented group of investigators of nature and bring them under the moral (and Christian) auspices of the philosopher.[14] That Huxley had ignored the English roots of the term indicates how unpopular the word had proven. Whewell's neologism was not widely adopted in the Victorian period, both because many practitioners had their own highly developed moral sense of vocation, which differed from Whewell's, and because practitioners resented the subordinate and restricted connotations of the term. But by the time Huxley registered his own protest against what he took to be a vulgar Americanism, the word "scientist" was taking hold in Britain. By the end of the century, that distinctly Victorian persona the "man of science" was beginning to vanish, and the "scientist," whose authority derived from laboratory discipline and from juxtaposition to literary culture, detachment from society, and disengagement of "facts" from "values," was already emerging.

---

[14] Ross 1962: 71–5. On Whewell's concerns to define science in relation to philosophy, see Yeo 1993: 32–8.

1

Science at Home

I have nearly traversed half the globe and have found only error and discord till I came to your cottage, where truth and happiness reside.
– Bernardin de St. Pierre, *The Indian Cottage*[1]

In 1846, Thomas Huxley received an appointment on HMS *Rattlesnake*, a survey vessel bound for the South Seas. In his shipboard diary, the twenty-one-year-old called himself a "man of science," but the designation was highly tenuous. His official title was assistant surgeon, a low-ranking officer in Her Majesty's Navy. With only two years of formal schooling, Huxley had been apprenticed to general medical practitioners in Coventry and London's East End. With the help of a scholarship, he had taken courses at Charing Cross Hospital and had read comparative anatomy and physiology in the library of the Royal College of Surgeons. Having completed the first examination for the degree of Bachelor of Medicine at University College, but lacking the financial means to continue his education, he entered the navy in 1845. A position on a survey voyage afforded a young man an excellent opportunity for furthering a career in science; however, Huxley was not the official naturalist on the *Rattlesnake*. This title fell to the ornithologist John MacGillivray, whose father was a professor of natural history at Aberdeen. Such dredging and dissection as Huxley desired to perform would have to be supplementary to his medical duties. His scientific findings were not guaranteed a place within the official report of the voyage.

[1] Bernardin de St. Pierre 1828: 288.

Huxley's status as a "man of science" thus was uncertain. But both the cultural identity and social role of scientific practitioners were themselves undergoing pronounced transformation at this time. A restructuring within learned societies and educational institutions, and the emergence of an ideology in which gentlemanly character could be acquired, alongside one in which it was innate, had together opened possibilities for young men like Huxley, the youngest son of a failed schoolmaster. Paid positions, however, were still scarce, and precisely what sort of community these men were entering remained unclear.[2] To obtain his naval post and subsequent appointment aboard the *Rattlesnake*, Huxley had to move in circles where patronage operated in tandem with meritocracy, yet where the criteria of merit were not firmly established. Like other aspiring scientific practitioners whom he would later befriend – William Carpenter, John Tyndall, Edward Forbes – Huxley presented himself as hard working, self-reliant, and averse to the entrenched privileges of aristocracy. But like them also, he positioned himself within a high-culture tradition whose bearers possessed inherent and lofty powers that raised them above other commercial and professional men.

In fashioning himself as a man of science, Huxley drew in part on models of genius he had gleaned from romantic literature.[3] He copied long citations from Carlyle's essay "Characteristics" into his diary while still a medical apprentice in 1842, passages that conveyed this romantic persona in some detail. According to Carlyle, genius dwelled in solitary minds whose sparks of thought, once kindled, could inspire action in the multitude. Genius was a heroic intellectual force, which swept the individual along as it carried the age, and yet remained mysterious even to its visionary self.[4] Conceived by eighteenth-century writers as an inborn and effortless capacity, genius lingered on in the Victorian period to describe a variety of self-made man fraught with contradictions: the genius-at-work whose peculiar labor was original rather than mechanical, moral rather than base.[5] Genius was innate, like nobility, yet it often resided in those of humble birth. It consisted partly of

---

[2] On scientific vocations in the first half of the nineteenth century, see Kargon 1977, Berman 1978, Cannon 1978, Morrell and Thackray 1981, and Inkster and Morrell eds. 1983. On the mid- and late-Victorian period, see Heyck 1982, J. Secord 1985 and 1986b, Schaffer 1988, and Gay 1997. The shifting meanings of "character" in the Victorian period are discussed in Collini 1991.

[3] On romantic models of genius, see Schaffer 1990. Other historical discussions of genius include Battersby 1989, Murray ed. 1989, and Shapin 1990.

[4] Carlyle 1831, quoted in T. H. Huxley Papers, Imperial College of Science, Technology, and Medicine Archives, London (hereafter HP): 31.169, "Thoughts and Doings", journal entry for April 1842.

[5] For eighteenth-century accounts of genius, see, for example, Gerard 1774.

characteristics such as intuition, mental suppleness, and refined dis-
crimination that the Victorians increasingly identified with feminine
nature. Yet they also held genius to be firm in its grasp and disciplined –
allegedly masculine qualities of mind.[6]

Huxley's character as a man of science thus slipped between Victorian
conventions of class and gender. In the early stages of his career, he utilized models of genius in conjuction with Victorian ideals of domesticity.
He presented himself as someone injured by the strife and self-interest
that governed public life and whose manhood depended on securing
a place of work that was removed – like the Victorian sanctuary of
"home" – from the sordid intrigue of politics and the grinding routine of
professional pursuits.[7] By identifying scientific work with the pure and
often feminized domestic sphere, he claimed moral distance from the
allegedly corrupt character of other forms of masculine, remunerative
work. Huxley's "man of science" was, fundamentally, a gender identity,
which entailed particular constructions of the home and of women.[8]

Despite the solitary nature of genius, and Huxley's own tendency to
brood while aboard the *Rattlesnake*, his scientific identity was formed not
in isolation, but through a process that involved the active contributions
of women.[9] Huxley met Henrietta Heathorn, the woman who would
eventually be his wife, while he was on shore leave in Sydney in 1847.
After four or five meetings over a period of six months, they became
engaged. Because of Huxley's difficulties in establishing himself as a
man of science after his return to England, the couple could not marry
until 1855, an extremely long engagement even by Victorian standards.
Over this eight-year period, in which Heathorn resided in Australia and
Huxley on board a surveying vessel and then in London, they exchanged
several hundred letters and kept journals for each other to read during
the long intervals of separation. Their correspondence was perhaps the
most important medium through which his identity as a man of science
and hers as a wife were shaped. Their protracted separation and arduous
social climb forced to the surface many of the assumptions about and
contradictions concerning gender during the Victorian period.

---

[6] The gendering of manners and mental characteristics in the eighteenth and early nineteenth centuries is examined in Outram 1989, Schiebinger 1989, Vincent-Buffault 1991,
and Barker-Benfield 1992.

[7] On the Victorian ideology of "separate spheres" of work and home, see, for example,
Houghton 1957 and Davidoff and Hall 1987. See also, however, Vickery 1993 and
Wahrman 1993, for critical accounts of historians who have taken this ideology as de-
scriptive, rather than prescriptive, of gender relations in the period.

[8] On Victorian masculinity, see especially Hilton 1989, Clarke 1991, and Tosh 1991.

[9] Works on gender and the sciences important in framing this account include Outram
1987, Jordanova 1989a and 1993, Daston 1992, and Goldstein 1994.

Imperial and Sentimental

Identity troubles appear early in Huxley's journals and correspondence written aboard the *Rattlesnake*. As surgeon-naturalist on a survey expedition venturing into uncharted waters and visiting unseen isles, Huxley could explore the dark interiors of little-known forms of marine life and make their contents his own.[10] Even before the ship's captain, Owen Stanley, began sounding the Torres Straits and naming islands and mountain peaks for himself, Huxley was working to affix his own name to the field of marine invertebrates, reclassifying Cuvier's Radiata and installing a new order of his own designation.[11] At sea, he had hoped to find an intimate, loyal community of scientifically minded fellows, an auspicious blend of culture and empire in which commerce and militarism were civil.[12] Just a few months after departing England, however, he described how his shipmates made fun of his books and threw his laborious dissections overboard as waste, while the captain remained aloof, apparently disrespectful of learning. Withdrawing from this rough fraternity, Huxley pined for the "social ease" and "friendly influences of a home circle." Above all, he longed for the fellowship of his sister Lizzie, her cultivated mind and taste, her "tenderest heart," and her "more than man's firmness and courage."[13] With both this sister, now emigrated to the United States, and Edward Forbes, a well-placed London naturalist who would become his chief patron, he located the trust and sympathy he missed on board the *Rattlesnake*. With them, he began to share his community of flora and fauna and his arduous search for zoological symmetries.[14]

When he was not dredging and dissecting, or reconstituting a domestic sphere through correspondence, Huxley had the company of novels. Many of these were of the sentimental genre and explicitly linked the occupations he plied in isolation on shipboard: the study of nature and the pursuit of hearth and home. Among the books that his coarse companions mocked were romantic tales about cultured men of feeling whose mission was to domesticate the world with truth. In Goethe's *Werther* and Carlyle's lives of Heine and Jean Paul, men of genius performed

---

[10] On the extensive utilization of imperial motifs by British naturalists during the period, see J. Secord 1982, Browne 1992, and Drayton 2000.

[11] Huxley's early research program, which was guided by a classification scheme known in contemporary zoological circles as Quinarianism, and his relationship with the chief author of that scheme, William Macleay, are carefully examined in Winsor 1976: 81–97.

[12] See especially his letter to his sister Lizzie, 6 October 1846, in L. Huxley ed. 1900, 1: 26–7.

[13] Diary entries for 10 January and 25 December 1847, in J. Huxley ed. 1936: 15, 71.

[14] On patronage relations as forms of domesticity, see especially Outram 1987.

through learning what women could achieve through feeling: the re-
finement of rough and rude nature and the softening of harsh men
whose public lives were devoted to struggle.[15]

In a letter to his mother from Mauritius, the setting of Bernardin de
St. Pierre's Rousseauist fable *Paul et Virginie*, Huxley expressed both
attraction and antipathy toward the ideals of sentimental fiction. "In
truth," he wrote, "it is a complete paradise, and if I had nothing better
to do, I should pick up some pretty French Eve (and there are plenty)
and turn Adam." Instead, he visited the tombs of two storybook lovers,
whose tale he believed to be "a fiction founded on fact": "Paul and
Virginia were at one time flesh and blood, and... their veritable dust
was buried at Pamplemousses in a spot... visited as classic ground."
The resting place was a garden wilderness; the lovers' ashes lay in two
funeral urns, each raised on a pedestal. Huxley made a sketch of the
scene and, returning with a pair of roses to scent his desk, was prompted
to remark, "I never was greatly given to the tender and sentimental, and
have not had any tendencies that way greatly increased by the elegancies
and courtesies of a midshipman's berth."[16]

Though Huxley was more at home with nature and novels than with
other agents of empire at sea, he could not be a sentimental culture hero
without a host of guilty associations. His berth filled with books, testimonies of domestic affection, and exhibitions (fragrant and foul) of
natural beauty, he had to insist to his mother and to himself that his sen-
timentalism had been extinguished by intercourse with a world ruled
by self-seeking and discord.[17] Epitomizing the middle-class Victorian
morality Huxley was espousing, Samuel Smiles would characterize the
home as a place of enlightenment and civility, "suitable for the growth
of the manliest natures," while criticizing writers like Rousseau and
Bernardin de St. Pierre as effete.[18] By mid-century in England, the pure
and regenerative ethos of the home had been reconstructed by several
generations of writers with the expressed object of bounding women's
domain. Men's work too was refashioned as a wilderness of strenuous
trial, the necessary complement to and practical support for domestic
bliss. Within this secular theology of separate spheres, the activities of
cultured men might be denigrated as ornamental, leisurely, or effeminate precisely because these men performed the social role consigned

---

[15] Diary entry for 24 December 1847, in J. Huxley ed. 1936: 70.

[16] L. Huxley ed. 1900, 1: 34. See also J. Huxley ed. 1936: 28–30. For a discussion of *Paul and Virginia* in relation to enduring associations of women with nature and men with culture, see Jordanova 1989a: 33–34.

[17] On shifting attitudes toward sentiments and sentimentality, see Outram 1989, Vincent-Buffault 1991, and Barker-Benfield 1992.

[18] Smiles 1871: 44–57.

to the home rather than engage in the muscular exploits of commercial and military men who profited by toil and stoic endurance.[19] Because of their own dependence on industry and the military for professional opportunities, men of science could not easily abjure these models. To gain a livelihood and position of moral authority as a man of science, Huxley would have to move between the spheres of work and home and between the host of gender opposites they implied. Dwelling as he did with novels, immersed in microscopic order and beauty, and seeking the learned community of his sister and Forbes, Huxley tried while on ship to construct a place of work that could be simultaneously a place of domesticity, albeit one constantly under siege by rude forces from without.

Though risky, such mingling of cultural tropes and activities widely considered to be distinct was precisely what would confer eminence upon men of science; for it was how other figures of high culture and moral gravity – men of letters, clergymen, captains of industry, even monarchs – were represented in a period when the home and the women within it were the bastions of everything pure. Much has been written about debates between these groups – almost exclusively groups of men – for cultural authority.[20] But the meaning of cultural practices like science, literature, and religion was of course not settled by men alone, nor were the women who participated in these settlements only those who had gained a public voice. Much negotiation over scientific identity took place between men and women in intimate conversations and correspondence. While Huxley was brooding over the significance of his science at sea, women were his chief respondents. Shortly after the *Rattlesnake* reached Australia in 1847, he met the woman he would eventually marry; she would replace his sister and mother as his principal confidante.

A Woman's Writing

While Huxley was sailing around the South Seas sketching the natural world, trying to make a name for himself as a man of science, Henrietta Heathorn lived in the home of her brother-in-law and half-sister outside Sydney, where she managed their household and helped raise their two children. She had lived in Australia since her middle teens, having emigrated from England by way of Germany, where she studied literature for several years. The details are sketchy, but it is clear that her father,

[19] The status of middle-class values of work and utility among early Victorian gentlemen of science are examined in Alborn 1996. On the masculine cult of industry more generally, see Collini 1989 and N. Clarke 1991.
[20] See especially Turner 1993.

Plate 1. Thomas Huxley in 1846 (from L. Huxley ed. 1900).

who owned a brewing business, was often insolvent. Most likely, it was
to ease her father's financial burden that Heathorn, at the age of sixteen or seventeen, went to live with her half-sister. Her brother-in-law,
William Fanning, was a successful merchant, and Heathorn assumed
considerable responsibility in running the Fanning home. She also accompanied the Fannings to parties and balls where she mingled with
the upper ranks of Sydney society. At one of these, she met Huxley and
shortly thereafter, at twenty years of age, became engaged to him.

A portrait of Heathorn as a devoted wife and as hostess to the great
has been consistently drawn by Huxley biographers and follows readily from her surviving documents, almost all of which were written for
Huxley's eyes. In a fragment composed near the end of her life, she
described first meeting "the young officer" at a dance, the interest they
found in their talk, and his calling on her later, astride a swift horse,
whereupon he paralyzed her with a fixed gaze and offered to remove
all hindrances from her path in life.[21] The poetry she wrote during their

[21] HP: 62.37.

Plate 2. Henrietta Heathorn (Courtesy of Virginia H. Huxley).

engagement and the letters and journal entries from the period recreate such chivalric bliss again and again: "I always knew some day the Prince would come for me. He blew the horn and stormed the gates and slew the giants in his way."[22] Like Huxley's airy zoological theories, however, these fairy stories were not conceived in complacency, but were often constructed as places of order and beauty where there seemed to be none. Expecting to marry after Huxley's return to London and promotion to full surgeon, Heathorn found her engagement prolonged for four additional years while her would-be husband shunned his medical duties to pursue a scientific career. Her own ideals of home and womanhood were repeatedly called into question by the aspirations of her fiancé toward forms of manhood and work that were in many respects opposed to prevailing models of masculinity and professionalism. Their physical separation, together with the considerable distance between their ideals, created serious conflicts. For Heathorn, writing was a process through which these conflicts could be reenacted

[22] H. Huxley 1913: 14.

and resolved on her own terms. It was perhaps her best means of over-
coming the discrepancy between their goals: "I have promised to keep a
journal and this promise made to one inexpressibly dear shall be faith-
fully kept – a journal not only of daily occurrences but thoughts which
bad or good shall be registered, even tho' intended for his perusal, for
should he not see me as I am? I will hide nothing from him."[23] Far from
being simple depictions of their relationship or of her own life and feelings, Heathorn's reverential and self-deprecating writings to Huxley
could serve her as a means of negotiating their differences and of
circumscribing spheres of gender and public and private space of
her own.

In the journal that Heathorn began the day Huxley departed for his
last survey expedition in 1848, she frequently described for him the
everyday activities that gave her greatest satisfaction. As manager of
her brother-in-law's household, she made the bulk of the purchases
for the home and commanded its servant staff, which included a cook,
butler, maid, nurse, and gardener. Her activities were typical of those
portrayed by historians of women as essential to the Victorian economy –
activities that Victorian women themselves, despite much middle-class
moralizing to the contrary, regarded as "work."[24] Heathorn defended
this position at the very outset of her relationship with Huxley. Not only
did she frequently refer to her household responsibilities as labour, she
did so against "gentlemen's assertions" that women were creatures of
leisure who lived to shop and against Huxley's own urging that she
spend more time "improving" herself: "The day passed as Mondays
generally do, very busy all the morning in household matters and all
the afternoon at work ... It is absurd of Hal to bid me read and practise
regularly – what with making my own things in the house I have full
employment."[25] An aspiring poet and devotee of German literature,
Heathorn too had visions of the house as a home for polite culture. But
she had a very different notion from Huxley of culture's place in the
home and of its role within her own economy of improvement.

Heathorn's domestic activity held much of its significance for her
as a form of religious devotion. Representations of the home as sanctuary and heaven, and of the wife as innocent angel and nurturing
madonna, were among the most important evangelical contributions to

---

[23] Diary entry for May 1849, in J. Huxley ed. 1936: 211. On the importance of private acts
of writing for Victorian gentlewomen, see especially Peterson 1989.
[24] On the central role of women in the Victorian middle-class economy, see Davidoff and
Hall 1987. For an account of the strenuous occupation of household management in
Britain, see Branca 1974.
[25] Diary entries for 26 May and 28 June 1849, in J. Huxley ed. 1936: 214, 218.

separate-spheres mythology.[26] Feminist writers have shown convincingly how these symbols reflected the social and political anxieties of ruling classes of men. Endowing women with religiosity because their yielding nature made them superior spiritual vessels was continuous with other constructions of passive, frail femininity.[27] But many women found in religious beliefs a moral meaning for their work in the home, as well as a moral foundation for their authority over others in it. If, as social historians have argued, religious institutions were places where Victorian women could exercise power and a range of expression that were not available to them in other spheres of public life, certainly one of the most important religious institutions was the household. Heathorn's own religious life, while including regular attendance at church, radiated from the place where she led the family in hymns at the piano, prayed for strength for her daily tasks, and cared for her half-sister's children with what she described as "holy love." Like her practical role of household manager, this devotional role received frequent articulation.

The discussion of religious matters between Heathorn and Huxley was in some ways typical of that between men of science and their often more orthodox wives or fiancées. Like Emma Wedgwood in her correspondence with Charles Darwin before their marriage, Heathorn gathered evidence for scriptural accounts of creation to parry her fiancé's skepticism and pointed out the uncertain foundations of all knowledge, including that of science.[28] More often, however, she proudly displayed the importance of Christian virtues for her command of the household. In a long passage from her journal she told how, returning home one evening to find the staff all "tipsy," the horse escaped, the kitchen on fire, and the nurse hurling abuse, she astonished a friend by her forceful manner of reproving the servants. In a measured tone, she quelled the inferno, restored order to chaos, and admonished the insolent creatures for intemperance: "I sent for the gardener to sleep in the house and shut up immediately that I might get them off to bed and after very quiet and firm measures restored the house to peace . . . they being penitent . . . I forgave them after very serious lectures."[29]

If the moral authority conferred upon middle-class women by religion was exercised most frequently over servants, it also legitimated

---

[26] On the centrality of religion in the home of the evangelical middle class and its practical effects for women, see Prochaska 1980, A. Owen 1987, and F. M. L. Thompson 1988: 250–3.

[27] For numerous examples, see Shuttleworth 1990.

[28] Burkhardt and Smith eds. 1985–2001, 2: 122–3, 126, 169.

[29] Diary entry for 23 October 1849, in J. Huxley ed. 1936: 276–8.

reproofs of men of the house whose conduct did not measure up to the standards of such women. It is uncertain if Heathorn expected Huxley to conform to a prevailing model of pious Victorian manhood – a husband and father who led the family in prayer, Bible reading, or religious instruction by the fireside and who took his place as head of the household in the parish church. But the religious doubts that he shared with her clearly rendered him a hindrance to her in her spiritual pursuits: "I fondly hoped," she wrote in her journal, "he would have been the guide and instructor unto more perfect ways – but here my hopes have borne bitter fruit. Something has come over me of late; I cannot pray as fervently as I did."[30] Such a confession may, however, testify less to Heathorn's dependence on Huxley for religious provision than to her own doubts about one who would make suspect the sacred center of her home. In extending her domestic and religious role through writing, she could intervene in matters where she felt Huxley was wanting or where she felt she had been wronged:

> The afternoon brought you, my own dear one... but you were in such a strange mood that I felt you did not make me glad. You were capricious; if I talked you would have me silent and if I laughed it grated on your ear and only at length, when I looked as sad as I felt and suggested I would try crying, did you utter any kind words or fold me lovingly to you... Before dinner time your fitfulness had returned and a little nasty spirit possessed me to tease you – till you warmly told me I had better not. I then of course felt more wishful than ever to do so and returned again and again... half in sport and half in earnest – till I went up to dress for dinner and told Alice part of my imaginary grievance. She had guessed something from your altered manner... [and] remarked you were dull and not joyous as usual. I reasoned myself into no very amiable mood much to Alice's amusement and so went down to dinner with a white dress but a naughty dark heart, punishing myself for your supposed harshness instead of you... Alice to whom I told all would not have me wretched all the evening and whispered [to] you I was unhappy – and then dearest you talked and kissed away all my evil fancies... How happily the rest of the evening passed, peace stole into my heart and abode there and when he had gone and I laid my head on my pillow I resolved that I would never more torment myself and him again. Love will tarnish if 'tis always petted.[31]

In the published version of Heathorn's journal, edited by her grandson Julian Huxley, it is suggested that this passage be read as evidence of her emotional volubility: under the strain of her ardent love for Huxley and his imminent departure, she retreats to a woman's world of taunting

[30] J. Huxley ed. 1936: 228–9.
[31] J. Huxley ed. 1936: 240–2.

and tears, where facts are indistinguishable from flights of fancy. But Heathorn's story was of course a retrospective one written for a man whom she considered capricious. Her tears, teasing, and woman friend were resources in what she staged as a contest of wills. In the tale, her friend's intervention resolved the conflict. But the process of writing itself produced a deeper resolution. In this scene and many others, it was always she who succumbed, she who could not make her needs understood. But with the closing moral and a change of person – here from "you" to "he" when referring to Huxley – she moved into a highly fictional voice and wrote a happy ending to the distressing evening. By such narrative shifts, Heathorn located Huxley in her literary frame-work. Thus writing enabled Heathorn to shape the meaning of her fiancé's actions and, in so doing, to make the ethics and politics of their relationship explicit, both to herself and to him. Through journal writ-ing, she developed the strategies she needed to cope with, and to act upon, their differences.

Such lively portraits of the young couple during their last few months together in Australia were drawn by Heathorn over and over again. In one incident, Huxley demanded that she give up her only photograph of him, which she had treasured for three years and had slept with under her pillow. He first promised to return it but then said that she could never have it back and that she was not even to ask for it.

> I felt he was asking almost more than I could perform ... surely he had done it on purpose to vex me – he was tyrannizing – he knew I could refuse him nothing and he asked so much – slowly and with difficulty I quelled the agony that filled my heart and promised not to ask again ... For a moment I felt he was unkind to try me – only for a moment – for I remembered how few ways I had to shew my love for him. I had yielded my will to his, even more a secondary affection; and I was, if not happy, at peace, for he would see that I would rather pluck the dearest object from my heart than offend him.[32]

In this passage, Heathorn mobilized the virtue that was typically demanded of men in public but was expected of women in private: sacrifice. By employing a traditional role, that of the self-effacing woman, Heathorn could transform the compromising of her will into an exercise of will. If Huxley, who later returned the photograph in a locket that she could wear around her neck, could not convey his love for her without displaying his power over her, Heathorn made his dominion an expression of her own self-command. In such narrated encounters, Heathorn described a relationship in which Huxley's

---

[32] Diary entry for 18 March 1850, in J. Huxley ed. 1936: 239.

affection was conveyed through assertion and hers through yielding, a relationship in which his gifts required that she give up something but hers were not received on her terms. She constructed her place in a world in which women had only very restricted means of expressing their worth, although the men they loved offered them fresh occasions for abnegation. Through her writing, she was making this virtue clear to a man who gave no indication that the terms of their relationship were troubling, who showed no signs of knowing the intricate ways in which her happiness depended on the movement of her will, and who, after denying her other ways of manifesting her devotion to him, might have failed to recognize even this – her self-denial. If an air of mystery and caprice, if self-command and a command of others, were the characteristics of genius, perhaps there was also something in Huxley's manner of the coarse masculinity that had troubled him about his shipmates and their abuse of military power.

> You draw out my thoughts and feelings – and appropriate them most tyrannically – and yet 'tis perhaps one of the things that has bound me with stronger love to you. You are a tyrant still conquering by strength where influence fails, indeed you have tonight acted very meanly ... and I have half – only half a mind, remember – to give you up as Will was constantly advising.[33]

While private accounts of emotion, unlike published or fictional accounts, are often read as literal, in this case it appears that Heathorn constructed her most intimate feelings and relations with Huxley as a literary text and that she actively employed the cultural resources available to her to model her life as one of sacrifice. In several instances, she reinterpreted the same literary material that Huxley had used on board ship to reconcile his life as a sentimental genius with that of a conquering hero. She read Carlyle's essay on Heine, which Huxley had used to console himself in his scholarly discontent, as a narrative of self-conquest through the subordination of one's own feelings to a greater existence and activity. Choosing this route to personal happiness and fulfillment – a route frequently taken by Victorian women who entered the public sphere – enabled her to manage the difficulties of her household duties through moral example.

> I rose a better creature, more cheerful and happy. He [Heine] struggled thro' deepest poverty and pain. Mind conquered the infirmities of the body and the evils of life ... And shall not I whose troubles are but faint and miniature shadows of his, strive against and subdue them? Henceforth I will ... All sorrow is selfish. I will become better and God help me in my intentions that they be deeds not words.[34]

---

[33] Diary entry for 26 March 1850, in J. Huxley ed. 1936: 242.
[34] Diary entry for 27 November 1849, J. Huxley ed. 1936: 286.

In prescribing for herself sacrifice before a man of learning, Heathorn performed her own elaborate cultural fabrication. Conventions of domestic sanctity and authority did not compel her, as they did some Victorian women, to make an insurgent entry into the public world of letters and politics. But they did enable her to assume the integrity of her household work, to transform literary texts and the power relations implicit in them through private acts of reading, and to compose subversive narratives to be read by her own (fallen) culture hero-cum-saint.

Improvement by Domestication

If Huxley, despite his learning, vocation, and invention, was still an unrefined and imperfect man, perhaps Heathorn could complete him. But according to whose model of manhood? In one of his first letters to his fiancée after their engagement, Huxley wrote that her confiding, tender love had awakened him to a nobler life and a purer course of action:

> Man is and must be influenced by the woman he loves . . . She is that living ideal of goodness before his eyes . . . when one is sick of the world, of its petty intrigues, its lesser and greater selfishness and dirt eating, when one is disposed to think that earnestness and truth and firm kind goodness have utterly disappeared from the earth, how great a blessing it is to feel assured that there is yet one in whom all these qualities live and so verily form a part of everyday life – out of the storybooks . . . man is as clay in the hands of the father – woman.[35]

Here Huxley described his entry into a symbolic order of Victorian work and home – the former a place of trial and atonement undergone individually and without regard for the feelings of others, the latter a place of salvation ruled by sympathy and affection. This Victorian home and its values functioned to complement, facilitate, and redeem his (still imagined) public life. Engagement with Heathorn enabled him to mobilize the virtues of this domestic sphere to anchor his ambitious self. From her loving base he could enter a world of intrigue, pursue a livelihood, and yet maintain a sense that he was working for something pure. While Heathorn was martyring herself on his behalf, he was setting his own conditions through writing to her, requiring that she fulfill various supportive roles if he was to succeed as a man of science. At times, the process was reciprocal:

[35] Letters of 6 and 15 February 1848, T. H. Huxley–Henrietta Heathorn Correspondence, Imperial College of Science, Technology, and Medicine Archives, London (hereafter HH): 7–8.

The thought that it is my duty to discipline myself for her sake...nerves
my better feelings – and often her image is my good genius, banishing
evil from my thoughts and actions...you have purified and sweetened the
very springs of my being which were before but waters of Marah, dark and
bitter were they. And strangely enough, too, not merely is your influence
powerful over my heart, but my intellect is stronger, my thoughts more
rapid, my energy less exhaustible, I never could acquire more rapidly or
reason more clearly.[36]

In this instance, Huxley's image of Heathorn as a moral example
corresponded with her own vision of herself as the spiritual center of the
household. There was an asymmetry to the lovers' accounts, however;
for while Huxley's pledge to work for Heathorn opened a space for
her to act, her power, unlike his, was never "tyrannical." Nor was the
genius that he claimed to draw from her one that she would ever be able
to manifest as her own. Huxley could place himself in awe of feminine
power when that power braced his own masculinity, when it enhanced
his own ability to work. Through odes to Heathorn's influence he tried to
order his affections in such a way that love was a consolation amid
struggles in the world, a source of satisfaction for desires that, unsated,
would fester and detract from his labor.

In resorting to such gender conventions, Huxley addressed the am-
biguities of his own identity as a "man of science." The qualities of
genius that elevated him above other men – its volatility and excess, its
intuition, its power of passion superior to will – were also qualities associated with feminine weakness. Accompanying his manly posturing
were journal entries and letters to Heathorn describing his "restlessness"
and "instability of temper." His thoughts were like his "strides up and
down this quaking deck." His intellect was "acute and quick rather than
grasping and deep."[37] He was less the man who fought vigorously in
the world, than the one who longed for the tender comforts of home:
"I have a woman's element in me," he wrote his sister. "I hate the in-
cessant struggle and toil to cut one another's throat among us men, and
I long to be able to meet with some one in whom I can place implicit
confidence, whose judgment I can respect, and yet who will not laugh at
my most foolish weaknesses, and in whose love I can forget all care."[38]
Though his own identity as a man of science would rest on a confla-
tion of separate spheres, and of masculine and feminine agencies, this
was not an identity that he could as yet sustain. Still a ship's surgeon
at sea, he could not rely on the established boundaries of home and

[36] Diary entry for 25 November 1847, in J. Huxley ed. 1936: 65–6.
[37] Letter of 17–18 January 1847, HH: 2–3, letter of 23 and 27 July 1848, HH: 36.
[38] Letter of 21 November 1850, in L. Huxley ed. 1900: 1: 61–2.

workplace should his genius appear effeminate. When he was unable to manage powers operating on him that were womanly, he had to evoke Heathorn's "strong natural intelligence" and her "firmness of a man" to explain her influence.[39]

This courtship dynamic was reenacted elsewhere. The cultural resources that enabled Huxley to exoticize Heathorn were increasingly brought to bear in his representations of the native peoples and terrain encountered by the *Rattlesnake*. While on the coast of New Guinea during the second year of his engagement, he sketched a landscape "lovely in the extreme" and fertile in natural resources unused by its idle inhabitants. The natives were "very civil" in character if not in accomplishment, "diminutive," "perfectly modest," and of "primitive simplicity and kind-heartedness."[40] By thus manufacturing objects of discovery to complement his conquering self, Huxley could graft his budding scientific identity onto the more proven manly stock of his shipmates. Adopting a discourse in which colonial others – their land, bodies, and culture – were made innocent, feminine, and ripe for tutelage, a ship's surgeon could establish an enlightened paternity of science over the domains of imagination, feeling, and tradition unschooled by reason.[41] By the time of his last survey expedition in 1849, he had outfitted himself (still privately) as a noble imperialist:

> There lies before us a vast continent – shut out from intercourse with the civilized world...and rich...in things rare and strange. The wild and noble rivers open wide their mouths inviting us to enter. All that is required is coolness, judgment, perseverance, to reap a rich harvest of knowledge and perhaps of more material profit...a little risk is also needful.[42]

But Huxley's role in the manly mission of empire was still more intricate than his. In journal entries he described the ship's economy as bound by routine, its men devoted only to pay, petty intrigues, and "the dreary business of charting." He privately reproached the crew members for cheating the natives in economic exchanges, and he rebuked his fellow officers for senseless violence. Most often, his conscience-raising took the form of mocking their gallantry. How intrepid was the

---

[39] L. Huxley ed. 1900, 1: 62.

[40] Diary entries for 19, 20, and 26 August 1849, in J. Huxley ed. 1936: 218–28.

[41] On European expeditions to the South Pacific, see Smith 1985. On conquest narratives, see especially Pratt 1992: 142–55. On medieval models of the virtuous conqueror and rescuer of women, see Girouard 1981. The role of gender, and of women's writing in particular, in the construction of colonial discourses are discussed in Blunt 1994 and S. Mills 1994.

[42] Diary entry for 31 August 1849, in J. Huxley ed. 1936: 166.

"brave Captain," "the little man" who would have all the natives bound
on the beach to satisfy him of the security of "his little body" or who
could be scared off by "old women, if at all shrewishly inclined"?[43] In
thus exercising his critical capacity, Huxley could prescribe the virtues
of the proper (English)man. By renouncing violence, he could be more
manly than men of war who took the lives of others while failing to
risk their own. But this was a peculiar form of manliness, one that ab-
jured force, that had sympathy for intimate communities of the weak
and for the simplicity, honesty, and peacefulness that made them so –
a manliness that preferred studying natives to putting them to work.
The conqueror of a feminine landscape, Huxley was also the critic of
empire. By extending the virtues of domesticity beyond the household,
his "man of science" could refine the savage manners of imperialists
as well as those of primitives and could make the colonies a happy
home.[44]

As an assistant surgeon, however, Huxley was not the ship's moral
conscience-in-residence. If he had been the official naturalist, he could
have written an authorized account of the voyage, and his disparaging
remarks could have found a public place.[45] Made privately, his judg-
ments served, instead, to defend his ambivalent manhood to himself,
and to Heathorn. By offering to redeem the world, namely its men, from
the crass worship of power and mammon, he was trying to command a
role that was in important ways continuous with hers. But Heathorn's
view of her home as an economy supported by worldly work was also
a challenge to his view of his work as a home apart from the world. Her
domestic role drew attention to the material foundations of his science.
How could he ask her to leave her home for his when he had none of
any substance to offer?

Pressing Points of Economy

After Huxley's last stay in Sydney in 1850, the couple did not see each
other for four years. In London, Huxley's difficulties in finding a place
for his science in the world of professions were more concrete than they

---

[43] Diary entries for 12 August, 5 September, and 12 December 1849, in J. Huxley ed. 1936:
212–13, 234–45, 260.

[44] On a closely related form of manliness being preached at the time by liberal Anglicans
such as Thomas Arnold, Frederick Maurice, and Charles Kingsley, see Vance 1985,
Hilton 1989, and Wee 1994. On the important role of domesticity in British colonial
discourse, see Poovey 1988: ch. 6 and Pratt 1992: 155–71.

[45] The crew's treachery was gently reprimanded in MacGillivray 1852, 1: 270–1. Huxley
would obtain a public forum by reviewing MacGillivray's work, see Huxley 1854a.

had been at sea. Shortly before reaching England, he wrote Heathorn that he no longer expected a promotion to full surgeon on the basis of his scientific work, but he did hope for a shore appointment and a grant to publish his research. He took up temporary lodgings with his brother in London but soon moved to a place where he could leave his books about, read and write in solitude, and greet the "great world only when necessary": "I have drawn the sword, but whether I am in truth to beat the giants and deliver my princess from the enchanted castle is yet to be seen." After just three months, he wrote that any attempt to live by a scientific pursuit was a farce, that he could earn distinction but not bread, and that he would sacrifice it all to be with her, away from the "buzz of the world" in a "quiet cottage" with only the prattle of children "about our knee."[46]

These dire sketches of his life in London were interspersed, however, with accounts of his new attachments with "immensely civil" men of the metropolitan elite, men who supported his work and whom he had come to call his "scientific friends."[47] Richard Owen, he wrote, would do anything in his power for him. Edward Forbes would move heaven and earth. In recommendations that he gathered in applying for a professorship at the University of Toronto, some of the most distinguished men in Britain's natural history world – Thomas Bell, William Sharpey, John Gray, Charles Darwin – testified to Huxley's industry and his powers of intellect and expression.[48] Though Huxley failed to obtain steady employment, his election to the fellowship of the Royal Society of London had been quick, and the society's gold medal soon followed. On receiving the medal in 1852, Huxley praised a community of fellows who were so open, honest, and free from the motives of personal interest that they could receive into their midst men whose efforts were truly original:

> The memoir for which you have done me the honor to award the Royal Medal today was printed by this Society during my absence in a remote corner of the world... far away from all sources of knowledge as to what was going on here... I sent my memoir away to you with a doubtful mind – I questioned whether the dove thus sent forth from my ark would find rest for the sole of its feet. But it has this day returned, and with... an olive branch and with a twig of the bay and a fruit from the garden... I trust I shall never forget the kindness and the aid I have received upon all hands from the men of science of our country.[49]

[46] 31 January 1851, HH: 135; 14 March 1851, HH: 140.

[47] 31 December 1851, HH: 177; letter to Lizzie, 21 November 1850, in L. Huxley ed. 1900, 1: 62–3.

[48] All of these testimonials are collected in HP: 31.68ff.

[49] HP: 31.139. Huxley refers to his paper "On the Anatomy and Affinities of the Family of the Medusae" (Huxley 1849).

On leave at half-pay from the Admiralty, Huxley supported himself by translating German zoology texts, working for museums, writing a quarterly science column and other miscellaneous pieces for the *Westminster Review*, and occasionally lecturing at the Royal Institution of Great Britain. He passed his evenings at boisterous dining clubs. Returning one night from the Red Lions Club, the fraternal supper order of the British Association for the Advancement of Science, he wrote to Heathorn, "I have at last tasted what it is to mingle with my fellows – to take my place in that society for which nature has fitted me . . . I can no longer rest where I once could have rested."[50]

Among the men whom he found to revere and trust implicitly was his "old hero," Sir John Richardson, a "man of men" to whom he was indebted for his *Rattlesnake* appointment.[51] He also admired Joseph Hooker, sure to succeed his father as director of the Royal Botanic Gardens at Kew, and seated at evening banquets beside his bride-to-be, the daughter of John Henslow, professor of botany at Cambridge.[52] Another of his "warmest friends" was John Tyndall, whose own career path in the physical sciences in many ways paralleled his. Later, when Huxley was appointed to lecture regularly at the Royal Institution, where Tyndall also taught, the latter would write, "we are now colleagues at home, and I can claim you as my scientific brother."[53] Chief among his new cohorts and paragons of purity was Edward Forbes, lecturer in natural history and paleontology at the School of Mines, and paleontologist to the Geological Survey prior to 1854. Forbes, Huxley confided to William Macleay, was "a man of letters and an artist, he has not merged the *man* in the man of science – he has sympathies for all, and an earnest, truth-seeking, thoroughly genial disposition which win for him your affection as well as your respect."[54] Though he had yet to obtain a permanent place among them, many of these men with whom Huxley associated were bachelors employed in the Geological Survey, an institution conceived by its director Henry De la Beche as a household with himself as father.[55]

Encouraged by his acceptance within this surrogate family of scientific peers, Huxley persisted in representing his vocation as lying outside the sphere of practical work where it could remain free of the self-interest, jealousy, and ambition that characterized his social climb – corruption that he could then blame for impeding him. A "hidden force"

---

[50] 7 November 1851, HH: 172.
[51] 7 November 1851, HH: 172.
[52] 7 December 1850, HH: 156.
[53] c. June 1855, in L. Huxley ed. 1900, 1: 126.
[54] 30 March 1851, in L. Huxley ed. 1900, 1: 94.
[55] See J. Secord 1986b: 239–41.

impelled him, a "sense of power and growing oneness with the great spirit of abstract truth."[56] Unvalued in the world of professions, with no position to afford him income for a house, Huxley still defended his dignity in the "house of experiment":

> Women often forget that men are essentially different from themselves, that a man's actions cannot and ought not to be exhausted within the circle of his affections as their own may rightly be...No woman who knows her true interests will ever begrudge the time her husband gives to...Science or Art. They are her best allies, for they all require earnestness and faith and fixity of purpose for their successful cultivation. In this pure sphere, the soul sickened and sceptical from intercourse with men meets truth face to face...It returns to the world purified and thence fitter to recognize the good in all shapes, fitter therefore to love, for that means to recognize purity and goodness.[57]

Huxley could assure Heathorn that his scientific affairs were not adulterous, because domestic values were identical with the values of science and other learned practices. He was like other men in needing the nourishment of the domestic sphere, while extending his aims and purposes beyond it. But his work was not like other men's – base and self-gratifying. When men of science left their loving wives, they entered a moral sanctum like that of their own home – a "pure sphere" where souls communed with truth. Domestic affection and its restorative virtues reproduced and reinforced these men's creative powers and moral endeavors: their role as discoverers and educators; their duty to civilize savage men and redeem civil society, to make the world a home by ordering the relations of classes, nations, and races as they ordered plants and animals, as their own households were ordered.

Remaining in Australia until her fiancé could obtain a living sufficient for them to marry, Heathorn now received Huxley's letters at the home of her parents, the Fannings having moved back to England shortly after Huxley's departure. There she again assumed responsibilities of management, offering occasional advice on matters of household economy to her father, whose money problems had worsened through speculating in the boom-and-bust business of gold. Because of her practical experience, vastly superior to Huxley's, in administering to the needs of a household, she was not consoled by his high-culture cottages in the air. The uncertainty of a home with Huxley and the pressures of her new circumstances combined to instill in her a fear of domestic ruin that he, a scientific knight-errant, in debt, with a moral aversion to

---

[56] 7 November 1851, HH: 172. The expression "house of experiment" derives from Shapin 1988.

[57] 9 July 1851, HH: 164.

the practical, could not abate. As the engagement continued and Huxley failed to procure a position, while persistently devaluing temporal ends, she began to press him with points of economy he would rather deny. When government positions opened in Australia during the gold rush, Heathorn cautiously suggested that he pursue a career that would hasten their being together. More frequently, she urged on his scientific person a more professional comportment. She did not accept his otherworldly concept of vocation, nor his hasty and ill-conceived resolves – which always came after his failure to get a particular scientific post – to take up brewing with her father or shopkeeping. In 1852, after he missed getting the professorship at the University of Toronto, she wrote:

> You talk of coming out – but not to practice your profession – will you let me say that I think I know you well – that you are unfitted by your previous habits from becoming a . . . storekeeper – your vocation does not lie in business matters . . . you have been so long employed in a particular and less practical branch of the profession. I mean less practical in a particular sense. That you might require to turn your attention especially to the requirements of practice before you commence on your own account.[58]

Such revelations from Heathorn – Huxley's station in an impractical part of the scientific profession, his inattention to practical requirements – were perhaps less bruising to Huxley's ego than her persistent attribution of his genius to acquired cultivation. While Huxley, to sustain his confidence, had continued to cite his whole life history as testimony that secret powers had intended him for science, Heathorn suggested instead that what had kept him apart from the ordinary influences of men, in isolation, poverty, and lack of sympathy, were simply his "habits" and past "employments." What determined him to a life of esoteric studies and her to a life of fiscal concern were his present "tastes and pursuits."[59] By contending with his genius – the mysterious force that he believed had fitted him for science, and that had fitted him to her – she was calling on him to work in a manner that he had always resisted:

> I . . . only think it necessary to possess a competency that your own hopes and pursuits may neither be crushed nor impeded . . . dearly as I love you and you know how well that is, it cannot blind me to the fact that many an otherwise happy home has been beggared in peace from the perpetual anxiety for the future.[60]

By presenting him with accounts of his scientific genius as a poor remunerative pursuit, and of himself as one inexperienced and disinclined

---

[58] 20 June 1852, HH: 211.
[59] 18, 20, and 25 January 1852, HH: 183.
[60] 20 June 1852, HH: 211.

toward practical affairs – in short, as a man ill-equipped to support a wife and family – Heathorn had done something almost unforgivable. Because her role was to support him in his battles with the world, her appeals were interpreted by him as slights to the gravity of his calling and as challenges to his manhood. His response was to redefine the home as a material possession and to represent the women in it as consumers of leisure and luxury and as maidens of sensual delight. In this taxonomy – made possible by his entry into a new domestic space among scientific men – the home and the affections therein were "earthly temptations" distracting him from his sacred calling. Unable to understand the moral motives of his vocation, Heathorn was ranked together with those men of the world who had no appreciation for "earnestness," men for whom work was just a matter of selfish "likes and dislikes":

> So far as I can judge, you do not seem to think that there is one course of life I ought to pursue rather than another – your own generous heart would have me follow that which I like best, but only because I like it best . . . My brothers understand ambition and profit. Fanning understands ambition. None, not even you . . . seem to comprehend the noble love of intellectual labour for its own sake.[61]

Faced with these acrid and accusing accounts, Heathorn reaffirmed her respect for Huxley's special calling while reminding him again that the domestic foundation on which his science rested was unstable. Confronting the prospect of a marriage in which her feelings and desires could not be expressed without questioning his prerogatives as a man and betraying her commitments as a woman, she refused to allow either the values that defined the home to be dictated to her or the importance of her role in the home to be diminished in order to dignify his work. Huxley had boasted of powers to tame chance, to discover the laws and solve the great problems of nature. She knew him as one who, despite his return to London, was still embarked on a global enterprise exploring dark continents, oceans, and life forms, opening paths to commerce and civilization. She accepted that their places in the world and their powers to shape it were different. But in the margin of the letter in which Huxley had accused her of not loving him and of not understanding the value of intellectual labor, she inscribed the words, "he wronged me." In ensuing correspondence, she evoked his own cherished transparency, threatening to draw a veil between them if he would not let her express her feelings. She recalled for him that their culture and circumstances had placed them in different though overlapping spheres:

[61] 10 February 1852, HH: 186.

Will you make no allowances for me remembering how different is my life to yours? mine a round of simple duties without one engrossing thought but you – yours a hard struggle I know dear Hal but still full of keen and pleasurable excitement, the hope of winning a name and a position...Whenever I have considered the years, the life time that were to make me your companion I have been with the pleasurable conviction that however far beyond mine in cultivation our tastes were naturally the same – and whatever sorrows from without we might have to encounter I yet believed our home would possess the attraction of unity and be bright with that sunshine of the heart where fullest sympathy and closely blended interests intermingle with a deep and earnest love such as I know mine is.[62]

Conclusion: Fairylands of Science

Early in 1855, Heathorn, despite poor health, set sail for England, and the couple were soon married: "I feel I have got home at last."[63] In June of the previous year, Huxley had obtained a lectureship at the School of Mines and a position with the Geological Survey, both jobs unexpectedly vacated by Forbes who left London for a chair at Edinburgh. On their honeymoon the couple stayed with two of Huxley's scientific friends, Frederick Dyster and George Busk, and their wives at Tenby. Still too weak to stand, Henrietta Huxley was waited on by her husband and carried daily by him to the seaside. On the Dysters' verandah, she made drawings and wrote descriptions as her husband dissected fauna gathered in a day's dredging with Busk. "I like it," she told Dr. Dyster, "I am proud to be associated with his work."[64] Although the duties of raising a large family would make it increasingly difficult, she continued to assist her husband in his work, translating German texts and preparing diagrams to accompany his lectures.[65] Late in life, she would reminisce, as she did of their romantic engagement, of her place in what now seemed to her to be the "fairy land of science":

When we married I had not the least idea of the true meaning of Science...I had only the dimmest idea, if any, of how a description of a marine creature should win for him fame – or help in any way to bring about his obtaining a position that would enable us to marry...he did not enlighten me.

But once we were married I began to learn. The whole atmosphere was scientific – his occupation, his friends, his books, the public lectures

---

[62] 10 September 1853, HH: 250.
[63] HP: 31.80.
[64] HP: 31.82.
[65] See Desmond 1998: 268, 450.

he gave that I attended. Little by little I saw and understood the great problems that underlay the dissection of even a fish or plant...of the interchange of force and heat, of the wonders of chemical changes. It was a revelation that ennobled the world I lived in...leading to undreamt of possibilities...I lived in a new world, full of strange facts, but facts that were like fairy tales.[66]

Henrietta's role as Huxley's scientific assistant was not atypical. Victorian wives often worked at home as their husband's translators, amanuenses, illustrators, counselors, and critics.[67] But precisely what this work meant to her, to Huxley, and to a wider Victorian public is difficult to determine. A history of the working association of the Victorian scientific husband and wife still needs to be written. If the relationship of Heathorn and Huxley is at all exemplary (a question that awaits a far more extensive investigation than is carried out here), the meaning of science in the home and the flow of influence and authority there were not matters that the scientific men of the house could easily control.

Two figures have dominated biographies of Victorian men of science: the heroic genius and, more recently, the rising professional. In such narratives, women have had little place except as supporters of men or as victims of exclusion. When not featured in romantic sagas of genius, Huxley has functioned as an archetype, even a caricature, of the rising professional; Heathorn, his fiancée for eight years and his wife for forty, has appeared as his constant "help and stay" or as the occasional, if resilient, victim of his irascible nature and emotional despotism.[68] Until recently, accounts of women's participation in the sciences have tended to reinforce these narratives by focusing on the exceptional achievements of a few and by taking as the norm of women's participation a professional practitioner whose vocation was increasingly removed from the domestic realm during the period.[69] These portraits owe much to mainstream Victorian scholarship, which represent the period as an age of separate public and private spheres and an era in which gender roles were likewise rigidly distinguished.[70] More recent studies, however, have shown some of the ways in which women played an

---

[66] HP: 62.2.

[67] The collaborative work of couples in the sciences has been explored in Abir-Am and Outram eds. 1987 and Pycior, Slack, and Abir-Am eds. 1996.

[68] For accounts of Huxley as a rising professional, see Introduction. For assessments of Heathorn's role, see L. Huxley ed. 1900, 1: 36–7, J. Huxley ed. 1936: 268–9, Jensen 1991: 23–44, and Desmond 1998. For a feminist critique of Huxley's career that emphasizes women's exclusion from the professions, see E. Richards 1989.

[69] On the juxtaposition of professional science to domestic life, see Abir-Am and Outram 1987. The literature on women in science is analogous in this respect to women's history; see the range of approaches outlined in Scott 1983.

[70] For the literature on separate spheres, see n. 7 above.

active role in shaping the meaning of science, and in shifting the criteria of scientific contribution, even when their scientific activity took place outside of the professional sphere defined by men as the proper home of science. As popular science writers, women could appropriate and substantially rework for large Victorian audiences the theories and discoveries of male practitioners, just as Henrietta Huxley transformed her husband's identity through her correspondence and journals and, eventually, through her published poems.[71] Though women were usually segregated, their presence and performance in laboratories and the field could unsettle or challenge prevailing modes of scientific conduct and participation.[72]

In her poems and reminiscences, written over the course of her marriage, Henrietta Huxley portrayed her association with science as uplifting. Its moral principles, embodied in her husband, raised her, she thought, to "a plane far higher than that of most people."[73] Such remarks not only implied an acceptance of the models of work and home that men of science were constructing for their own much-needed fulfillment and assurance, they rested on an elaborate model of acceptance that had framed Heathorn's relationship with Huxley from the beginning. Every point in their engagement at which she had contended with him and found dignity in sacrifice depended on views of women's work as reproductive rather than creative and on views of the proper roles of women as instructors of children and the poor, popularizers rather than discoverers of knowledge, and dutiful attendants or silent muses to men of genius. Consoled by his wife, her independent identity henceforth invisible to the historian (except perhaps for her regular appearances at church and as Huxley's hostess),[74] Huxley went on to father highly moral scales of being and development that located and ranked all living things and to engage in running arguments about who should participate in constructing these scales and in defining these moral terms. By locating inferiority in women's bodies, minds, and daily tasks, the fields of natural history, political economy, and, just-emerging, psychology, biology, and anthropology, created by Huxley and other men, contributed to a social order in which women played a subordinate role.[75]

If Huxley's later cultural polemics and pronouncements did much to construct the image of the explicitly gendered "man" of science of the

---

[71] On women science writers, see Gates and Shteir eds. 1997.

[72] See, for example, Gould 1997 and Winter 1998b.

[73] H. Huxley, "Reminiscences," HP: 62.2.

[74] The extensive correspondence between Huxley and Heathorn after their marriage remains in private hands; however, an edition is currently under preparation (A. Darwin and Desmond, forthcoming).

[75] See E. Richards 1983 and 1997 and Russett 1989.

second half of the nineteenth century, this chapter has shown precisely how gender was negotiated in the construction of that image and why it is important to consider women in that negotiation. This process of gendering was not always reciprocal and mutually fulfilling. But amid considerable conflict and misunderstanding, what Huxley and Heathorn were able to give each other was surprising. She helped to transform his science into remunerative work. He made her home into a fairyland of science. Heathorn came to revere her husband as one whose soul glowed with purest fire, who "grasped the torch of God's own flame"; yet her adoration did not necessarily accord with his own ambitions. In her poems, and in her everyday activities in the household, she converted science into natural theology, into romance, into industry, even into part of a woman's work at home.[76] It was precisely this sort of conversion – achieved in "popular" and "amateur" practice – that the activities of "professional" practitioners like Huxley were designed to limit or dismiss.[77] The terms in which scientific identity was settled by Heathorn – as a useful profession, as a support for domestic economy, as a form of religious devotion, and as a literary art – suggest the variety of meanings science held for its middle-class publics and the degree to which the discourses and authority of science were negotiable by other social groups. Huxley's efforts to establish and control the identity of science for others, both in his own profession and in the spheres of literature, religion, and politics, will form the subject of the ensuing chapters.

[76] See H. Huxley 1913: 11–12, 75–6, 99–101, 155–6, 158.

[77] On the mutual construction of the "popular" or "amateur" and the "professional," see Cooter and Pumphrey 1994, A. Secord 1994a, and Winter 1998a. Wide-ranging reflections on resistance in everyday practice may be found in Certeau 1984.

2

Gentlemen of Science?

Debates over Manners and Institutions

I have just received H's coarse-looking little book [*Man's Place in Nature*] –
not fit as somebody said to me, for a gentlemans table.[1]
                                    – Joseph Hooker to Charles Darwin, 1863

As Huxley sought a scientific career, first at sea and then in London, during the late 1840s and early 1850s, he drew on a range of models of manliness from the imperial culture of exploration and conquest and from the
heroes of sentimental fiction. He also appropriated ideals of Victorian
womanhood and domesticity to distance himself from and obtain moral
authority over other forms of commercial and industrial endeavor.
Isolated for much of this period from the metropolitan world of learning, he conferred a social meaning upon his scientific work through the
novels he read, the journal he kept, and the extensive correspondence he
undertook with his fiancée. How then did Huxley conduct himself with
other gentleman practitioners whose company he now wished to join?
How did he position himself within this diverse scientific community?
Who became his models, mentors, and patrons?

[1] Hooker to Darwin, 26 February 1863. In Burkhardt and Smith eds. 1985–2001, 11: 179.

At mid-century, the sciences in Britain had little of the career structure and few of the defining institutions of today, such as the large research laboratory with its team of experts, the academic department, or the university degree.[2] For the metropolitan elite, many of whose members came from the aristocracy, the gentry, or the highest ranks of the learned professions (medicine, the clergy, the bar), scientific life was dominated by private societies, both general and specialist. Conduct in these societies was governed by rules of gentlemanliness. Admittance was not conferred only on the basis of expertise but also on the possession of appropriate manners, typically manners acquired through family connections and education at an elite school and university.[3]

Gentlemanliness was not simply a requirement for entry into elite scientific society, however. Rather, the practice of science within this society was itself gentlemanly – that is, science was a particular way (though a precarious one) in which a man in the Victorian period could become a gentleman. It was not necessary to be from the upper classes to enter this society. Some practitioners, like William Whewell and Richard Owen, rose from trade and craft backgrounds to the upper classes through their scientific vocations. However, because positions of scientific distinction were poorly paid, and paid positions in the sciences were still very few, it was difficult for those who lacked family and university connections, or the requisite wealth, to follow the scientific path to gentlemanliness. To do so required almost continuous negotiations within a complex network of patronage.

Once he had gained a foothold at the School of Mines in 1854, Huxley rose rapidly within the inner circles of metropolitan science. In the following year, he became both Fullerian Professor at the Royal Institution and lecturer at St. Thomas's Hospital, London. In 1856, he was made an examiner at the University of London and a fellow of the Geological and the Zoological Societies. He later acquired the Hunterian Professorship at the Royal College of Surgeons, served as president of the British Association, and as secretary, and then president, of both the Geological Society and the Royal Society. Older biographies have tended to describe Huxley's success as a triumph of disinterested science over the social impediments of religious prejudice and gentlemanly favoritism.[4] In more recent accounts, his advancement is attributed to his position within a new, reforming, middle-class generation of professionals who swept away the aristocratic, old-boy system that blocked the career path of

---

[2] On metropolitan scientific institutions and social mores in the early Victorian period, see Becker 1874, Berman 1978, Morrell and Thackray 1981, Rudwick 1985, J. Secord 1986a, and Morus, Schaffer, and Secord 1992.

[3] On codes of gentlemanliness, see St. George 1993 and Morgan 1994.

[4] Bibby 1959, Irvine 1959, and Jensen 1991.

scientific researchers.[5] But if early Victorians practiced science in accordance with gentlemanly codes and patronage, this is because these social structures were held to nurture, not to impede, knowledge. The emergence of a "pure science" that might be corrupted by gentlemanly codes and patronage is itself historically contingent, and must be explained. The alternative argument, that Huxley's campaigns against tradition emanated from middle-class, professional interests, is also difficult to sustain. Huxley's first mentor and role model, Edward Forbes, though a salaried employee of the state, moved comfortably within a world of patronage. Huxley's foremost scientific hero, Darwin, was the epitome of the gentlemanly specialist, while his chief antagonist, Richard Owen, was one of England's most prominent scientific professionals and a figurehead for many reformers in government and in medicine.

Yet over the course of his career, Huxley was able to substantially reshape the public image of the "man of science," particularly that of Darwin and Owen. Drawing on a range of identities, Huxley presented practices still current in the sciences, including politeness and patronage, as social obstacles and aristocratic holdovers. Recent works have explored the ways in which manners, as expressed in day-to-day relations between philosophers and men of letters, helped to constitute learned communities in the seventeenth and eighteenth centuries.[6] If such approaches seem inappropriate for the second half of the nineteenth century, this is because scholars have accepted the accounts of professionalizers like Huxley at face value – namely, that "manners" no longer mattered. During the Victorian period writers such as Anthony Trollope, Robert Browning, John Stuart Mill, and Carlyle introduced new codes of frankness and honesty framed against existing conventions of politeness.[7] Precisely how these general critiques of gentlemanly civility shaped, and borrowed from, scientific modes of expression, address, and comportment deserves much more attention.

Huxley's efforts to reform scientific identity also operated on an institutional level. There were in fact competing professional models for the life sciences in Victorian Britain: one was oriented around museum display and public instruction and the other around laboratory research and specialist training.[8] Although Huxley moved between these different domains for most of his career, from early on he sought to establish

---

[5] Turner 1993 and Desmond 1998.

[6] Shapin 1994 and Goldgar 1995.

[7] St. George 1993: 226–32.

[8] On the development of museums in the nineteenth century, see Stocking ed. 1985, Sheets-Pyenson 1988, Jordanova 1989b, Forgan 1994, and Yanni 1999. On the museum and laboratory contexts of Huxley's work, see especially Forgan and Gooday 1996 and Desmond 2001.

the laboratory as the preeminent site for the production and dissemination of knowledge. Huxley's protracted disputes with Owen, whose prominence and celebrity were closely tied to those of the natural history museum, repeatedly raised the question of which institutional settings were most conducive to proper scientific conduct and the advance of knowledge.

The Survey Man

After his return to London in 1850, Huxley found his ideal of the man of science embodied in Edward Forbes. A decade earlier, when Forbes himself had been an aspiring young naturalist, there was no professional norm in Britain for pursuing the sciences.[9] But in many respects Forbes's career path became exemplary for naturalists in the mid-Victorian period. For Forbes, private reading followed by a university course, an apprenticeship in the field, and then, in 1841, an appointment as naturalist on HMS *Beacon* bound for the Mediterranean were steps that provided the broad natural historical training and detailed mastery of a particular domain essential for assuming a leading place in the natural historical community in Britain.[10]

On his return to Britain, Forbes faced a situation similar to that which would confront Huxley after the *Rattlesnake* voyage ten years later. After field training and survey work, the career options of naturalists diverged widely. Those of independent means, like Darwin, could continue to work from the retirement of their homes. For others, medical practice or clerical office had long provided bases for natural historical research. In the first half of the nineteenth century, the frequency and magnitude of expeditionary voyages provided opportunities for some, like John Richardson, to combine scientific work and naval careers. Still others, like Edward Blythe and Hugh Falconer, obtained positions as botanical administrators or curators in the colonial territories. Alfred Russel Wallace earned his living through travel writing and the sale of specimens.[11]

Academic positions in the metropolis for naturalists were scarce. The value of Forbes as a model for Huxley lay partly in the fact that Forbes had succeeded in obtaining a number of institutional posts and thus represented a new type of institutionally based researcher and teacher.

[9] On professional opportunities in the chemical and physical sciences, see Hays 1983, Neve 1983, Smith and Wise 1989: chs. 2–4, and Cantor 1991: ch. 6.

[10] On Forbes, see Rehbock 1979, E. Mills 1984, Gardiner 1993, and Preece and Killeen 1995.

[11] On Richardson, Blythe, and Falconer, see Stephen and Lee eds. 1885–1912. On Wallace, see Raby 2001.

In 1842, Forbes acquired his first appointment in London, the chair in botany at King's College. He immediately supplemented this position with one as lecturer and curator at the Geological Society. In 1844 he became lecturer in natural history and paleontology at the School of Mines, and paleontologist to the Geological Survey – both being institutions housed in the Museum of Practical Geology on Jermyn Street. As an employee of the School of Mines and Geological Survey, Forbes was part of a team of geologists and paleontologists hired by the government in the mid-1840s when the survey and museum were brought under the jurisdiction of the Office of Woods and Forests and a comprehensive minerological map of the British Isles was planned. This group of survey men, headed by Roderick Murchison, was an early example of the kind of professional scientific community that would become increasingly prominent in the second half of the century: a group of largely anonymous, paid researchers organized according to a division of labor and bureaucratic chain of command and promising benefits to the expanding industries of Britain.[12]

At the same time, this new institutional structure was tied to the sort of imperialist enterprise of discovery upon which Huxley had portrayed himself as engaged in his *Rattlesnake* diary. In his 1854 review of Murchison's *Siluria*, Forbes praised the author's stratification system for giving a British stamp to the world's rocks. He also commended the generosity with which Murchison acknowledged the others whose work had gone into developing the system. Here the model of scientific community was patently militarist and imperialist. Murchison, whose *Siluria* drew upon the researches of a dozen survey men, was portrayed as commander in chief, a heroic discoverer leading his "noble army of investigators" with "the energy of fifty hammers."[13] Allied with these manly qualities, Murchison's "genius" and "love of truth" marked him out as an exceptional, original thinker amid his company of men, giving him the power to command others, as well as to order nature. This romantic and military model reveals some of the complexity of the scientific community in which Forbes was operating. The heroic genius could reside alongside the new anonymous scientific practitioner, who was simultaneously a soldier and a technician. Institutions such as the Geological Survey, the Royal Observatory at Greenwich, and, later, the Cavendish Laboratory at Cambridge, all of which were devoted to the production of universal standards and measures, were simultaneously expressions of British superiority and centers of individual and

---

[12] J. Secord 1986b.
[13] Forbes 1855: 25–6, 35. The imperial culture of Victorian natural history is discussed in J. Secord 1982, Stafford 1989, Browne 1992, and Drayton 2000.

ineffable invention.[14] This institutional structure appealed to Huxley because it seemed to be more impersonal and meritocratic than the patronage system that had long regulated scientific careers. The authority and prestige that Forbes gained from his institutional position did not compromise his independence of mind.

Codes of patronage, politeness, and deference characteristic of the gentlemanly tradition of British science lingered on among these new professionals, however. Huxley had been introduced to Forbes and a number of other metropolitan naturalists by Owen Stanley shortly before the departure of the *Rattlesnake* for the South Seas. His prominent position in the London natural history community, along with his research in the field of marine invertebrates, made Forbes a most valuable contact for Huxley. Huxley wrote to Forbes at length of his work on the Portuguese man-of-war and its implications for the reclassification of the Radiata (see Chapter 1). He received no reply and learned only on his return to London three years later that his work had met with a favorable reception. With Forbes's assistance, Huxley's article, "On the Anatomy and Affinities of the Family of the Medusae," posted off to England in 1848, had been published in the Royal Society's *Philosophical Transactions*.[15] Shortly after his return, Huxley began to pay regular visits to the Museum of Practical Geology and was given a key to Forbes's office. In November 1851 he wrote to William Macleay, describing Forbes as "a first-rate man, thoroughly in earnest and disinterested, and ready to give his time and influence – which is great – to help any man who is working for the cause."[16] Earlier that year, Huxley had written his sister that Forbes had supervised papers published in his absence, provided many important introductions, helped him in getting membership to the Royal Society, and penned a favorable review of his work – "all done in such a way as not to oppress one or give one any feelings of patronage... I can reverence such a man and yet respect myself."[17]

Everything that Forbes did to help Huxley build his career – introductions, lobbying for election, laudatory reviews – was in keeping with the tradition of patronage. It was only Forbes's departure in 1854 from the School of Mines to fill the natural history chair at Edinburgh, formerly occupied by Robert Jameson, that gave Huxley the permanent institutional position he had been seeking for four years. Yet Forbes was able to exercise his influence on Huxley's behalf in ways that Huxley

---

[14] On the Royal Observatory under George Airy, see Schaffer 1988. On the Cavendish Laboratory under James Clerk Maxwell, see Schaffer 1992 and 1995.

[15] Huxley 1849.

[16] Huxley to Macleay, 9 November 1851, in L. Huxley ed. 1900, 1: 102.

[17] Huxley to Elizabeth, 20 May 1851, in L. Huxley ed. 1900, 1: 103–4.

found acceptable. What Huxley singled out for admiration in Forbes was his manner. Forbes fit his ideal of the man of science because he seemed to respect the work of others, however different from his own, and because he never presented himself as Huxley's superior. Forbes's manner of conducting himself with Huxley enabled the latter to build a sense of himself as an autonomous practitioner, gaining through hard work and originality. But Forbes's mode of associating with his own superiors was quite different. It allowed figures like Murchison to appear as warrior-chieftains or kings, specially endowed with the ability to create order and to establish principles for others to follow. In effect, the scientific community to which Forbes belonged was a mix of democracy and patriarchy. In it, progress could result from either humble and anonymous industry or the innate, masculine powers of the mind.

Huxley's relationship with Forbes was more complex than his glowing character sketches let on. In his letters to Forbes from the *Rattlesnake*, Huxley had presented himself with both confidence and deference. He was a humble fact-gatherer, offering his unusual finds as rare gifts to the expert theoretician, and an original thinker, able to formulate new ideas on invertebrate classification that would reorganize Cuvier's Radiata.[18] For Huxley, Forbes was a useful person to know because he was deeply embedded in a network of patronage and hierarchy, and it was by using this network that Huxley could gain his professional position. But Forbes was not the only man of science whose image and patronage Huxley needed to cultivate. Forbes's unexpected departure to Edinburgh in 1854, and his sudden death a year later, made Huxley's relations with another metropolitan naturalist, whose manner Huxley did not admire, both more significant and more contentious than they might otherwise have been.

The British Cuvier

Though Forbes was well known and influential within natural historical circles in Britain, most contemporaries would not have chosen him to exemplify the scientific practitioner in the 1840s and 1850s. For many, that person was Richard Owen. Despite Huxley's subsequent glorification of Darwin as a scientific role model, Owen was in many respects closer to the man of science that Huxley was trying to become. The eminent position that Owen had achieved by the 1850s was the result of various negotiations of the sort that Huxley too had to undertake. Moreover, Owen was highly supportive of both Forbes and Huxley, using his own

---

[18] Huxley to Forbes, September 1847, HP: 16.154.

position of influence to enhance their careers. In short, Owen's identity as a man of science was far from that of the petulant, conservative
opponent of Darwinian theory sketched in many accounts.[19]

Like Forbes and Huxley, Owen was compelled to move between social worlds to gain positions and influence. But the course he adopted
was quite different. Owen was born in 1805 in Lancaster, the son of a
West India merchant. His family background in trade actually placed
him socially beneath the sons of professionals, like Huxley, who composed most of what Desmond has called the "young guard" proponents
of "proletarian" science.[20] After medical training in Edinburgh, Owen
became a member of the Royal College of Surgeons in London and set up
a medical practice in Lincoln's Inn, hoping to attract young lawyers as
his patients. But from the late 1820s, he began to pursue a natural historical career in tandem with medicine, taking opportunities on both sides.
He became active in the Zoological Society of London and cultivated
ties with distinguished members of the natural historical community
such as William Buckland, Adam Sedgwick, and Charles Lyell. In 1827
he became engaged to the daughter of William Clift, his superior at
the Hunterian Museum; it was an engagement that, like Huxley's, was
extremely prolonged. Despite a promotion and two salary increases at
the museum, Owen did not have a livelihood sufficient to support his
fiancée until 1834, when the Royal College conferred upon him a new
chair in comparative anatomy at St. Bartholomew's Hospital. In 1837,
when he was made the first permanent Hunterian Professor at the museum, Owen was able to let his medical practice lapse. His chief duties
then involved the dissection and cataloguing of the many specimens be-
queathed to the museum by the physician John Hunter. His fame would
be founded upon his position at the Hunterian.

Many of Owen's patrons at the Hunterian were trying to draw on
the natural sciences to reshape medical education. But Owen's public
identity as a man of science was ultimately forged in opposition to the
medical establishment. In July 1837, he wrote to his wife Caroline concerning a request he had received from the college trustees to report at
once on his previous year's work:

> What's the use of trying to collect one's ideas for a report to the Trustees?
> "One thousand and three moths killed by tobacco-smoke and directions of
> the Board of Curators. Complaint of some of the sorrowing relatives of said
> moths that returns was used instead of canaster (such infra digs. would
> never have taken place in good old Sir William's time, the moths – though

---

[19] See for example, Hull 1973: 213–25, Bowlby 1990: 352, 450, and Desmond 1982: ch. 1,
72–4. A large literature is cited in Rupke 1994: 220, n. 3.

[20] Desmond 1982: 13.

they be moths – having been bred and born in the Royal College of Surgeons)." Secondly, "All old corners and out-of-the-way archives diligently and carefully looked through, and the letters, out o'date, old catalogues, and other documents, left where they were found." Thirdly, Mr. O. has minutely and casually looked (without spectacles) at all the uncatalogued specimens in spirit, and feels much out of spirits himself when he thinks of the same. Fourthly, that Mrs. O. closed the due proportion of her windows after the demise of his late most gracious Majesty and Patron of the College, and also wore mourning no less becoming to herself than to the melancholy occasion. Fifthly, Mrs. O.'s kitchen chimney still smokes, contrary to the directions of the *late* Chairman of the Board and the wishes of the Trustees. I cannot get on; it's no use.[21]

Unlike many of the gentlemanly practitioners with whom he associated, Owen had no private fortune or estate. His scientific and domestic existence consisted of his small salary and the provision of living quarters at the Hunterian. His livelihood depended upon a time-consuming and often unrewarding daily chore of classifying someone else's specimens. Owen's career path was thus a considerable gamble. For a young man of his background, medicine was a secure and established profession, capable of yielding a comfortable and respectable lifestyle. A life of science, however, was not a self-evidently gentlemanly pursuit for those who had no private income. Owen has often been portrayed as an establishment figure, a benefactor of aristocratic patronage who used his influence to impede the progress of rising young professionals – especially Huxley – all of whom had to make their way in opposition to the system in which Owen was entrenched. But Owen was every bit as much a self-made man as Huxley and others of this younger generation – perhaps even more so. Owen's position as a scientific practitioner in the Hunterian was almost unprecedented. It was, in fact, a new vocation in the scientific world in Britain.

In making his own way professionally, Owen did have one prominent model available to him, one that he would labor throughout his life to install on equal terms within Britain. The model in question, the natural history museum, was well established on the Continent, epitomized perhaps by the Paris Muséum d'Histoire Naturelle, which Owen had visited on several occasions. The institutional structure of museums in Britain was built almost entirely during Owen's lifetime, with the new British Museum of Natural History opening in South Kensington just a few years before his death. Nearly two hundred new museums were founded in Britain during Victoria's reign alone, exhibiting a wide range of objects from the fine arts, the sciences, fashion, and industry. Recent work suggests that Owen was the individual who did the most

---

[21] R. S. Owen ed. 1894, 1: 116–17.

to shape the museum in Britain as a site for the employment of scientific practitioners, often in the face of great resistance.[22]

Owen made a name for himself in the museum world by dealing with the curious or rare animals such as the platypus and the gorilla that most fascinated the British public. His reconstructions of giant extinct reptiles and mammals from fossil remains won him a reputation as a naturalist in the mold of Georges Cuvier. Owen's Hunterian lectures were widely reported in the *Lancet*.[23] His popularity was akin to that of Humphry Davy, whose regular demonstrations at the Royal Institution became media events attracting fashionable audiences.[24] As such, Owen contributed to the public status of the man of science as a public figure, dispensing natural knowledge to audiences seeking rational improvement and entertainment. Owen became an active member of some of the most exclusive London societies, including the Athenaeum and The Club. At posh dinner parties he discoursed like a great hunter on his "big bird" – the moa.[25] But he was also an active participant in the Red Lions Club, the dining order set up by Forbes and others as an alternative to the expensive formal gatherings at British Association meetings. From the mid-1840s, the Red Lions met regularly in London for food, drink, and song and became a social center for young metropolitan naturalists, including Huxley.[26] Owen also became a figurehead for reformers who were concerned in different ways to enroll the sciences in the improvement of higher education, industry, and government. With other scientific practitioners like Henry De la Beche and Lyon Playfair, Owen served on a series of prestigious metropolitan commissions to improve the health of towns, the condition of sewers, and the London meat supply. Owen was unique among naturalists, however, in coming to represent the man of science as deserving of public support.[27] From 1840, he began to receive regular invitations to dine with the prime minister, Robert Peel, who soon obtained for him a Civil List pension of two hundred pounds to assist his scientific career, offered him a knighthood (which Owen for the present declined), and commissioned a full portrait of Owen, to stand opposite that of Cuvier, in Peel's private gallery at Drayton.

Owen was perhaps most at home in the day-to-day activities described matter-of-factly in his wife's diary: lecturing to a crowded Hunterian theater, finishing early because a trustee fell asleep, and retiring to a large dinner party in the museum rooms, followed by two Corelli

---

[22] Rupke 1994: chs. 1 and 2.

[23] Rupke 1994: 17, 28, 91–92.

[24] On Davy as a public figure, see Berman 1978 and Knight 1996.

[25] R. S. Owen ed. 1894, 1: 238.

[26] On the Red Lions Club, see Gardiner 1993 and H. and J. Gay, 1997.

[27] On the national endowment of science in Britain, see MacLeod 1970a, 1971a, and 1971b, and Alter 1987.

sonatas with himself on violincello, or attending a meeting of the Geological Society, and then eating a meal at home with Buckland and the paleontologist Gideon Mantell, followed by auto-experimentation with microscope and blood:

> Dr. Mantell, who stated that he had a very slow circulation, on examination proved to have blood globules of a decidedly larger size than the others. Dr. Buckland was just saying with that droll look of his, "Why, Mantell, you see you have a good deal of the reptile about you," when the news was brought in that the Queen was safely delivered of a little princess, so the discussion was stopped by all the gentlemen drinking health to Her Majesty.[28]

An anatomical pieceworker, a civil servant in an industrial metropolis, a benefactor of the Crown, a chevalier to the king of Prussia – Owen's identity as a scientific practitioner was as complex and controversial as his circle of patrons, friends, and employers. His wide-ranging social relations and negotiations give the lie to the traditional view of Owen as exemplary of the aristocratic "old guard" and show the inadequacy of "class" as an analytical category for his career path as a naturalist, his public role as a man of science, or his relations with gentlemanly naturalists and government figures. Most of the associations that furthered and structured his success were with particular kinds of reformers seeking institutional status for the sciences on a par with their status on the Continent, and enlistment of the sciences in a large-scale reorganization of British institutions. By 1850 it seems that many of the career difficulties that the next generation of scientific practitioners would face had already been surmounted by Owen. At the Jermyn Street museum, first Forbes and later Huxley would encounter problems analogous to Owen's at the Hunterian. They would struggle to advance their scientific vocations in an institution originally built to display mineral resources and to teach new engineering professionals. By raising the status of natural history in elite society, in imperial culture, in progressive, state-sponsored reform, and in popular education and amusement, Owen helped prepare strategies that later naturalists could capitalize on.

As Huxley's early relations with him show, Owen used his established position to help others in ways that no one else could. When Huxley returned from his survey expedition on the *Rattlesnake* in 1850, he tried to obtain a shore appointment and a leave of absence with half-pay so that he could write up his research of the last four years into a monograph.[29]

---

[28] R. S. Owen ed. 1894, 1: 177.

[29] *The Oceanic Hydrozoa* (T. H. Huxley 1859a) was eventually published by the Ray Society, an organization founded to support the publication of scientific works that were not commercially viable (Curle 1954).

He approached Owen, whose connections in government made him invaluable in such matters, to assist him in these negotiations. Owen's testimonial on Huxley's behalf was one of a series he provided over the next two years, including letters of support for natural history posts that opened in Toronto and Aberdeen. In November 1852, Huxley asked Owen for another reference – this to be sent to the secretary of state – in support of a government grant for his research. When the letter did not come after several weeks, Huxley wrote to Forbes, who attributed the delay to the "physiological condition of Owen's memory."[30] Huxley asked Owen again, and again received no reply. He then wrote to Forbes describing the following encounter with Owen on the street:

> I then on the fourth day afterwards, met him. I was going to walk past, but he stopped me, and in the blandest and most gracious manner said, "I have received your note. I shall grant it." The phrase and the implied condescension were quite "touching" – so much that if I stopped a moment longer I must knock him into the gutter. I therefore bowed and walked off. This was last Saturday. Nothing came on Monday or on Tuesday, but on Wednesday morning I received "with Prof. Owen's best wishes," the strongest and kindest testimonial any man could possibly wish for! I could not have dictated a better. I immediately sent a copy to Mr. Walpole.
>
> Now is not this a most incomprehensible proceeding?
>
> I gave up any attempt to comprehend him from this time forth.[31]

Forbes's response was sympathetic, but he urged conciliation:

> He is certainly one of the oddest beings I ever came across and seems as if he was constantly attended by two spiritual policemen, the one from the upper regions and the other from the lower – the one pulling him towards good impulses and the other towards evil. As I believe men's bad qualities in 3 out of 5 instances are generated in the stomach I lay many of O-s various eccentricities to the charge of his ill health. He has very much that is good and kind in him, with all his faults.[32]

Despite Forbes's apparent efforts at peacekeeping, from very early on Huxley began to present Owen as a behind-the-scenes manipulator. Convinced in 1852 that Owen would try to prevent the publication of his article "On the Morphology of the Cephalous Mollusca," he wrote to his fiancée that Owen considered "the Natural World as his special preserve and 'no poachers allowed.'"[33] But Owen's letter to Huxley from March 1853 stated that he had received Huxley's article with "great pleasure" as "additional evidence of the scientific activity of one who works so much

---

[30] Forbes to Huxley, 16 November 1852, HP: 16.170.
[31] Huxley to Forbes, 27 November 1852, HP: 16.172.
[32] Forbes to Huxley, 2 December 1852, HP: 16.174.
[33] Huxley 1853. Huxley to Henrietta Heathorn, 5 March 1852, L. Huxley ed. 1900, 1: 105.

after my own heart."[34] Owen had in fact done nothing to impede the publication of Huxley's work and was experiencing the same difficulties as Huxley in publishing his own technical manuscripts.[35] Huxley's slights to Owen's character continued apace with his attacks on Owen's choice projects. Ironically, it was Huxley's harsh 1854 review of the latest edition of *Vestiges of the Natural History of Creation*, in which he ranked Owen with the still anonymous author as proponents of the progressive development of species – a theory "unworthy of serious attention" – that turned Owen on the offensive.[36] In the following year, Owen's new edition of the *Comparative Anatomy and Physiology of the Invertebrate Animals* omitted any reference to Huxley's published criticisms of Owen's theory of the alternation of generations. It also included a remark on Huxley's "blindness" regarding the brachiopod heart, which involved the neglect of Owen's own work.[37] In turn, Huxley's 1856 review of the same book highlighted Owen's "disposition to exalt himself at [others'] expense."[38]

Various explanations have been offered for the protracted controversy between Owen and Huxley, ranging from differences of personality, to competing philosophical systems (Owen's idealism, Huxley's materialism), to class conflict.[39] Combining these interpretations, Desmond argues that it was Owen's arrogance, on the one hand, and his socially conservative anatomy, on the other, that turned the combative, plebian Huxley against him.[40] Nicolaas Rupke's recent and extensive work on Owen also emphasizes the clash of personalities, adding a fair measure of professional rivalry. In place of Desmond's account of Owen as an ideologue of the British old regime, Rupke presents Owen as caught between conservative and liberal patrons, producing natural historical theories and museum projects to satisfy both.[41] Even when taken together, however, these analyses cannot encompass the variety of uses to which Owen and Huxley put science and the wide range of

---

[34] Owen to Huxley, 15 March 1853, HP: 23.249.

[35] On Owen's difficulties in obtaining a grant from the government or the Royal Society to publish his work on the Megatherium in 1853, see Desmond 1982: 28–9.

[36] T. H. Huxley 1854b: 5–6. The early disputes between Huxley and Owen are discussed in Desmond 1982 (see especially pp. 212–13 n.), and Di Gregorio 1984.

[37] R. Owen 1855: 493.

[38] T. H. Huxley 1856a: 27.

[39] The importance of personality has been emphasized by Bibby 1959: 72 and Ruse 1979: 143–4. Irvine 1959: 39–41 roots their antagonism in religious grounds. Paradis 1978: 118–21 and Di Gregorio 1984: 186–8, argue that Huxley's attacks on Owen were derived chiefly from his rejection of Owen's transcendental anatomy. Di Gregorio suggests that Huxley's professional ambitions also fueled the controversy. Lyons 1997 places the controversy within a context of religious and ideological differences, while resting Huxley's case against Owen largely on empirical grounds.

[40] See, for example, Desmond 1982: 38–41 and 1998: 177.

[41] Rupke 1994: 40–60.

values and movements that each man came to symbolize. Personal differences and divergent theories and political agendas clearly existed between a great many scientific practitioners during the Victorian period. Additional categories are needed then to understand both the conflict between Huxley and Owen and the eventual consolidation of scientific identity and community across generations, across class boundaries, and amid a host of theoretical differences and social disputes. As Rupke's research shows, the question of personality cannot be settled through a transparent reading of contemporary accounts. Against any number of derogatory portraits of Owen, many others can be marshaled to testify to his good nature.[42] More importantly, the focus on personality, ideology, and class conflict has obscured what, for contemporaries, was perhaps the central issue in the Huxley-Owen controversy – the "good character" of the man of science.

In his 1853 letter to Forbes, Huxley indicated that it was Owen's "manner" and the "implied condescension" that provoked him. Owen's manner was gracious, polite, and patronizing. Owen operated in a social world in which it was proper to be patronizing, a world in which deference and condescension were part of a system of manners within a learned community of unequals. Forbes too was rooted in this world, and he did not view this behavior as a discredit to Owen's character. Instead, he tried to mediate between the two men by suggesting to Huxley that Owen was not responsible for his manner. Although Huxley did not find Forbes to be patronizing, Forbes's glowing tributes to men of scientific renown like Murchison show that deference and condescension still had an important place in the manners of the new professional practitioners. When Darwin came to write his autobiography in 1876, he depicted Owen as his "bitter enemy" and attributed their falling out to Owen's "jealousy" over his success.[43] To contemporaries in the 1850s and 1860s, however, it was neither straightforward nor obvious that Owen's conduct was ungentlemanly. That Huxley's view of Owen came to predominate in accounts written toward the end of century thus needs to be explained. In effect, Huxley's "victory" over Owen was the triumph of a new code of scientific conduct. In the drafting of this code, the image of another practioner – Charles Darwin – was paramount.

The "Genius"

In December 1859, Huxley began his review of Darwin's *Origin of Species* for *Macmillan's Magazine* by praising the author's sterling scientific

---

[42] See the examples in Rupke 1994: 6–10.
[43] Barlow ed. 1958: 104–5.

character: "It has long been my privilege to enjoy Mr. Darwin's friendship, and to profit by corresponding with him, and by, to some extent, becoming acquainted with the workings of his singularly original and well-stored mind."[44] In this and other early reviews of the *Origin*, Huxley's assessment of the book's arguments was reserved and equivocal, his account of the book's author undivided and glowing. In his article for the *Times* in the same month, Huxley stated that the *Origin* could not fail to attract readers, "so clear is the author's thought, so outspoken his conviction, so honest and fair the candid expression of his doubts."[45] As the evolutionary controversies drew on, Darwin's virtues grew increasingly prodigious under Huxley's pen. In an 1871 review, he wrote that it was not only Darwin's "industry, his knowledge, or even the surprising fertility of his inventive genius" that struck the student of nature, "but that unswerving truthfulness and honesty which never permit him to hide a weak place, or gloss over a difficulty."[46] In a memorial printed in *Nature* in 1882, Huxley declared Darwin to have possessed "an intellect which had no superior . . . a character which was even nobler than the intellect . . . [and] a wonderfully genial, simple, and generous nature . . . The more one knew of him, the more he seemed the incorporated ideal of a man of science."[47]

In some respects, Darwin seems an unlikely figurehead for the professionalizing Huxley. He was a friend of John Henslow, Adam Sedgwick, and others in the "old-boy" circle at Oxford and Cambridge. Until the early 1860s, he was also a friend of Owen. He was independently wealthy, a country gentleman, and thus precisely of the social rank that, according to many accounts, was swept aside by the rising middle or "mercantile" class for whom Huxley spoke. Huxley could never imitate Darwin, whereas he could follow the path carved out by Owen. Huxley's relationship with Darwin was full of tensions that Huxley left out of his accounts and eulogies. Each had strong social and theoretical reservations about the other. Their friendship and collegiality were sustained at least in part through a considerable reshaping of each other's public image.

Huxley's friendship with Darwin was conducted, as were many of Darwin's relationships after he moved from London to Down, largely through correspondence. From 1851 to 1856 the two men exchanged a series of letters on Cirripedia, the barnacles about which Darwin would publish two lengthy monographs. They traded specimens for

[44] T. H. Huxley 1859b: 142.
[45] T. H. Huxley 1859c: 14–15.
[46] T. H. Huxley 1871a: 184.
[47] T. H. Huxley 1882a.

dissection and exchanged views on the cement glands, which Darwin believed were modifed ovaria. Their letters, in which knowledge was shared, credit given, and obligations incurred and fulfilled, indicate how a friendship could be established between scientific men over quite technical matters.[48] Also during this period, Huxley approached Darwin for testimonials, and Darwin consulted Huxley for references to German anatomy and morphology.[49] Huxley's working knowledge of German was extremely useful to Darwin, who read the language very slowly and with great difficulty, a considerable handicap for a professional physiologist and comparative anatomist.

The value of Darwin for Huxley as a patron is less clear. Though well known and highly regarded among naturalists, Darwin was not institutionally established like Forbes or Owen and could not command the influence in government or in metropolitan societies that Huxley needed to obtain either a position or funds to publish his research. Of the nine testimonials that Huxley gathered for his application to Toronto, Darwin's was the most laconic:

> I have much pleasure in expressing my opinion, from the high character of your published contributions to science, and from the course of your studies during your long voyage, that you are excellently qualified for a professorship in natural history. You have my best wishes in your present application.[50]

Darwin's letter of reference did not issue from an institutional base or a professional position, nor did Darwin have such resources to offer the young candidate. His support was delivered in a personal letter, a fashion that Huxley could not qualify as patronizing, like Owen's. In effect, Darwin gave his support in the same manner as he and Huxley conducted their private and friendly correspondence, in which differences of age, experience, and social status were not evoked – a gentlemanly letter written to a known gentleman.[51] Throughout their scientific correspondence, Darwin also repeatedly diminished his role as a patron or mentor to Huxley by claiming to be insufficiently knowledgable to comment on Huxley's comparative anatomy of marine invertebrates. What was conveyed in his letters to Huxley, even those in which Darwin functioned as a patron, was their shared position in a gentlemanly scientific community.

---

[48] For the correspondence of Darwin and Huxley between 1851 and 1856, see Burkhardt and Smith eds. 1985–2001, vols. 5–6.

[49] Darwin to Huxley, 17 July 1851, and 11 April 1853, in Burkhardt and Smith eds. 1985–2001, 5: 49, 130–1.

[50] Darwin to Huxley, 9 October 1851, in Burkhardt and Smith eds. 1985–2001, 5: 64.

[51] On the codes of gentlemanly correspondence, see A. Secord 1994.

The basis of this community was neither stable nor certain, however. On many occasions, the relationship between the men was strained by differences that were both social and theoretical. In 1856, Joseph Hooker wrote to Darwin about Huxley's nomination for election to the prestigious London club, the Athenaeum. Huxley had recently been awarded the Royal Society's gold medal, and his election would have helped to advance the status of men of science among the gentlemanly elite of the metropolis.[52] In reply, however, Darwin expressed concern about the possibility of an opposing faction, perhaps led by Owen, who would cite Huxley's dismissive attacks against Louis Agassiz and Cuvier in recent lectures.[53] It was not enough, Darwin stated, to have published some excellent papers in the Royal Society's *Philosophical Transactions*, for "scientific merit" alone was insufficient.[54] To nominate Huxley and fail would be detrimental to their cause. In a further letter to Hooker, Darwin wrote that while Huxley's lectures seemed clever and correct,

> yet I think his tone very much too vehement, and I have ventured to say so in a note to Huxley. I had not thought of these Lectures in relation to the Athenaeum, but I am inclined to agree with you and that we had better pause before anything is said. It might be urged as a real objection the way our friend falls foul of every one (N.B. I found Falconer very indignant at the manner in which Huxley treated Cuvier in his Royal Institution Lecture, and I have gently told Huxley so). I think we had better do nothing.[55]

Huxley was not put up for election that year, nor the next. His public criticism of the most eminent men in natural history was an insult to the tradition of patronage and the codes of deference, authority, and distinction that still governed elite Victorian society. Such unsociability prevented him, on this occasion, from gaining a position of great cachet and influence. Those who were most supportive of him in private were unwilling to act in public. The risk that his public persona posed to their own reputations in polite society was too great.[56] Huxley's personal assaults and public airing of controversies were regarded as inappropriate behavior not only by Darwin but by a whole generation of practitioners trained in the early Victorian period, when personal disputes or controversies over theory were not typically presented in the published literature. Such disagreements were deemed more appropriate in

---

[52] Hooker to Darwin, 7 May 1856, in Burkhardt and Smith eds. 1985–2001, 6: 103.
[53] For his criticism of Agassiz, see T. H. Huxley 1855; on Cuvier, see T. H. Huxley 1856b.
[54] Darwin to Hooker, 9 May 1856, in Burkhardt and Smith eds. 1985–2001, 6: 106.
[55] Darwin to Hooker, 21 May 1856, in Burkhardt and Smith eds. 1985–2001, 6: 111–12.
[56] Huxley's eventual admittance to the Athenaeum in 1858 is discussed in Chapter 3.

private conversation and correspondence.[57] To Darwin and many of his colleagues, Huxley's manners were not mere personal idiosyncracies, but forms of behavior that could jeopardize his standing in the gentlemanly world of the sciences. Darwin had a very different set of codes for conducting himself professionally, and he was evidently ambivalent about Huxley's manners in the 1850s.

In addition to his reservations about Huxley's scientific conduct, Darwin also disagreed significantly with Huxley on development. Although Darwin disqualified himself from critiquing much of Huxley's technical work, he was less reserved when the subject turned to progressivism or transformism. He expressed "surprise" at Huxley's opposition to anamorphism in the latter's 1853 article "On the Morphology of the Cephalous Mollusca," writing to Huxley in April of that year, "I should have thought that the archetype in imagination was always in some degree embryonic, and therefore capable and generally undergoing further development."[58] Darwin's surprise turned to disappointment the following year on reading Huxley's condemnation of Agassiz's theory of progressive embryological stages.[59] From the early 1850s, Huxley had used his lectures and articles to attack every version of progressivism of note. After acquiring a position as lecturer at the School of Mines in 1854, Huxley continued his attacks in the courses he taught and pitched them more aggressively, and more particularly, at Owen. Almost until their publication, Darwin kept his working hypotheses and manuscript on species development concealed from Huxley. In letters to Huxley, Darwin had distanced himself from Owen's transcendental use of the archetype. But with respect both to developmentalism in general and to codes of scientific conduct, Darwin was in fact much closer to Owen than he was to Huxley. In his own correspondence with Darwin, Owen had shown himself to be polite about, and perhaps even supportive of, Darwin's work on evolution. On receipt of his personal copy of the *Origin of Species*, Owen wrote, "For the application of your rare gifts to the solution of this supreme question I shall ever feel my very great indebtedness."[60]

Relations between Darwin and Owen extended back to the 1830s, when the Darwins lived in London, and the men and their wives met occasionally for dinner. Owen had received many specimens collected by Darwin from the *Beagle* and was awarded the Geological Society's

---

[57] The protocols governing scientific disagreement during the early Victorian period are discussed in Rudwick 1985: 24–7 and J. Secord 1986a: 21–3, 171–2.

[58] Huxley 1853. Darwin to Huxley, 23 April 1853, in Burkhardt and Smith eds. 1985–2001, 5: 133–4.

[59] Darwin to Huxley, 2 September 1854, in Burkhardt and Smith eds. 1985–2001, 5: 212–13.

[60] Owen to Darwin, 12 November 1859, in Burkhardt and Smith eds. 1985–2001, 7: 373–4.

Wollaston medal in 1838 for his work on some of Darwin's fossils. In turn, Owen read proofs of Darwin's narrative of the *Beagle*'s voyage and wrote positive reviews of Darwin's work. Such gentlemanly support continued after the publication of Darwin's evolutionary views. In his Presidential Address at the British Association meeting in 1858, Owen gave a favorable account of the papers of Darwin and Wallace on natural selection.[61] In November 1859, he wrote to assure Darwin that such a theory was not "heterodox" to him, for he had himself long been "disposed to believe in the operation of existing influences or causes in the 'ordained becoming and incoming of living species'" and had been gravely rebuked for it.[62] In December of the same year, Darwin wrote to thank Owen for the intended present of Owen's forthcoming book, an edition of Hunter's *Essays*, which included this "precious note" on the species issue: "The best attempt to answer this supreme question in zoology has been made by Charles Darwin in his work entitled 'The Origin of Species.'"[63]

Rupke's work has shown that with respect to the transcendental nature of the archetype, and the issues of transformism and natural law, Owen's work had long been ambiguous. As Owen himself indicated to Darwin, he had received stern criticism for appearing to favor transformism. His efforts to arrange for an English translation of Lorenz Oken's *Lehrbuch der Naturphilosophie* had provoked controversy among members of the Ray Society.[64] He was admonished by Adam Sedgwick for his 1849 lecture "On the Nature of Limbs."[65] He had been ranked with the author of *Vestiges* in the 1854 review by Huxley and likewise condemned. Darwin and Owen may have differed over the causes of evolution, but their correspondence, together with Owen's public statements, suggest that the two men had a well-established mode of conducting such disagreements privately and cordially, while continuing to support each other publicly.[66] Huxley, however, not only was opposed to any theory of transmutation, he had attacked distinguished men of science, in public lectures and in print, for advocating such theories.

---

[61] R. Owen 1858a: xci–xcii.

[62] Owen to Darwin, 12 November 1859, in Burkhardt and Smith eds. 1985–2000, 7: 373–4.

[63] R. Owen ed. 1861, 1: 37; Darwin to Owen, 12 December 1859, in Burkhardt and Smith eds, 1985–2000: 7: 424.

[64] The translation, *Elements of Physiophilosophy*, appeared in 1847 as one of the Ray Society's official publications. For a discussion of the controversy surrounding the translation, see Rupke 1994: 230–3.

[65] R. Owen 1849. Sedwick 1850: ccxiv.

[66] See, for example, Darwin's letter to Lyell, 10 December 1859, in which he gives an account of a long and heated discussion with Owen shortly after the publication of *Origin of Species*, in Burkhardt and Smith eds. 1985–2001, 7: 421–3.

Instituting Biology

By examining the social and theoretical relations among Darwin, Huxley, and Owen before the evolutionary controversy over *Origin* – and prior to the polarized and simplified accounts of the factions involved – the stakes of that controversy can be differently interpreted. The men's disputes over theory were transformed through protracted debates over manners and institutions. Prior to publication of the *Origin*, Huxley's attacks on Owen and on other prominent natural historians like Agassiz tended to center on the progressivist or transformist features of their work. After the appearance of Darwin's *Origin*, Huxley maintained his reservations regarding evolutionary theory, but he began to change tack. On reading the *Origin* in November 1859, he wrote to inform Darwin that he was "sharpening up [his] claws and beak." The manners that Darwin had found objectionable in Huxley would now prove advantageous to Darwin: "Some of your friends are endowed with an amount of combativeness which (though you have often and justly rebuked it) may stand you in good stead."[67]

From the outset, Huxley's reviews of the *Origin* divided the Victorian world. On one side stood Darwin, endowed with "the calm spirit of the philosopher," and his sympathizers, who numbered "every philosophical thinker" and "all competent naturalists and physiologists." On the other side stood Darwin's critics, namely, "pietists, bigots, [and] old ladies of both sexes." All the critics were held up as proponents of "special creation" – the theory that presumed that species arise from a "supernatural creative act," and that owed its existence and plausibility entirely to "Hebrew cosmogony."[68] In Huxley's accounts, men of science all appeared to line up as "supporters of Darwin"; all others were in various ways unscientific, refusing to admit naturalistic explanations for the development of life. Despite Owen's sympathies, expressed privately and publicly, for Darwin's theory; despite the ambiguity of his work with respect to transformism, "natural" laws, and causes; and despite his own persistent campaigns against scriptural geology and literalist interpretations of Genesis, Owen was implicitly classed with those individuals who were either outside or on the margins of the scientific community, dismissed as obscurantists, bibliolaters, or ignorant laypersons.[69]

However, Owen was neither outside nor marginal, but central to the reshaping of British natural history institutions. As a result of the

---

[67] Huxley to Darwin 23 November 1859, in Burkhardt and Smith eds. 1985–2001, 7: 390–1.

[68] T. H. Huxley 1860: 22–3, 53–4.

[69] In his own review of *Origin*, Owen maintained that he held "no sympathy whatever with Biblical objectors to creation by law" (R. Owen 1860: 511).

enormous popularity of his Hunterian lectures, the surgeon had risen to the ranks of the gentry. When he left the Hunterian in 1856, he toured provincial towns, lecturing on local fossils and ruminants, and returned to London to begin a series of courses at the Museum of Practical Geology. It was after Owen had taken up this position at Jermyn Street, where Huxley had only recently been appointed, that relations between the two men reached a breaking point. Like his Hunterian courses, Owen's lectures at the museum were media events. It was with Owen at the lectern, speaking before a packed audience on some of his giant mammals, that the Jermyn Street institution received notoriety in the *Illustrated Times* in April 1857 (see Plate 3). According to the reporter, "the intelligent, mild-eyed Professor" was like a great hunter and "walks among the beasts with as much power – of a greater kind. His menagerie exhalts the mind, and astonishes the understanding."[70] When Owen, on the basis of his visiting lectureship, had his name entered in the 1857 edition of John Churchill's *Medical Directory* as "Professor of Comparative Anatomy and Palaeontology" at Jermyn Street, Huxley wrote the publisher insisting that Owen held "no appointment whatever in the Govt. School of Mines" and that Owen's entry was "calculated to do [Huxley] injury."[71] According to Huxley, Owen had sought to multiply his own appointments and influence by encroaching on Huxley's institutional position.

The image that Huxley began in the mid-1850s to fabricate for Owen, the "British Cuvier," as ambitious and power hungry, was similar to the tactic that Geoffroy Saint-Hilaire had used against Cuvier himself in the 1830s.[72] Huxley turned institutional power and political connections, which Owen had in spectacular degree, into evidence of corruption. At the end of 1856, Huxley broke all personal contact with Owen, writing to Frederick Dyster, "I would as soon acknowledge a man who had attempted to obtain my money on false pretences."[73] Subsequent sallies against Owen repeatedly impugned his manners as inappropriate to the man of science. For Huxley, Owen had come to signify a prisoner of the world of patronage. Owen thus could not be an original thinker, but was a "servile follower" who expected all others to be the same.[74] His commitments were not to truth, but to the maintenance of his position and against any person who might encroach upon it with contrary views.

---

[70] *The Illustrated Times* 4 (18 April 1857): 251–2. Owen's paleontology lectures at the Royal School of Mines were given from 1857 to 1861.

[71] Huxley to Churchill, 22 January 1857, HP 12.194.

[72] See Appel 1987, and Outram 1984.

[73] Huxley to Dyster, December 1856, HP 15.80.

[74] Huxley to Darwin, 3 October 1857, in Burkhardt and Smith eds. 1985–2001, 6: 461–2.

PROFESSOR OWEN LECTURING AT THE MUSEUM OF PRACTICAL GEOLOGY.

Plate 3. Richard Owen lecturing on the osteology and paleontology of mammals at the Museum of Practical Geology, Jermyn Street (*Illustrated London Times*, 18 April 1857).

Though the theoretical terms had changed, the same form of character assault was made by Huxley after the publication of the *Origin*, as the long dispute over brain anatomy illustrates. In an 1858 paper at the Linnean Society, Owen had proposed that humans be distinguished from apes, partly on the grounds of a structure unique to their brains, the hippocampus minor, thus creating a new subclass, the Archencephala.[75] Huxley quickly objected to this, arguing for the continuity between humans and apes in his Royal Institution lecture "On the Special Peculiarities of Man."[76] By 1861 the debate had become a public showcase for Huxley's attacks on Owen's manners and merit. At the British Association meeting of that year, George Rolleston, newly elected Linacre Professor of Anatomy at Oxford, read from a letter by Huxley that charged Owen with an "obstinate reiteration of erroneous assertions."[77] In the same year, Huxley established his own journal, the *Natural History Review*, the first volume of which carried his article "On the Zoological Relations of Man with the Lower Animals," together with supporting papers on the orangutan by Rolleston and William Church, Lee's Reader in Anatomy at Christ Church. Huxley's piece devoted eight pages to a refutation of Owen's work, building his case on the dissections of chimpanzee brains performed by his friends Allen Thomson and John Marshall, the latter a surgeon at University College Hospital. Specifically referring to his letter read by Rolleston before the British Association earlier that year, Huxley staked his own integrity and reputation on the outcome of the dispute: "essential as I have felt it to be to my personal and scientific character to prove that my public assertions are entirely borne out by facts, I am far from desiring to deal with this important matter in a merely controversial spirit."[78]

Throughout the long and increasingly public and embittered controversy over the existence and interpretation of anatomical "facts," Huxley never raised the issue of what theories, evolutionary or otherwise, might be implied by those facts. Rather, his persistent concern was with the implications that the controversy held for the proper behavior of men of science. On learning that Owen was up for election to the Royal Society Council in 1862, Huxley wrote several letters to the secretary, William Sharpey, insisting that "one of us two is guilty of wilful and deliberate falsehood" and that therefore Owen's election turned on the question of "whether any body of gentlemen should admit within itself a person who can be shown to have reiterated statements which are false and which he must know to be false." Claiming that a favorable vote

---

[75] Owen 1858b.

[76] The manuscript of the unpublished lecture, delivered on 16 March 1858, is in HP 36.96–108.

[77] As reported in the *Athenaeum*, 21 September 1861, p. 378.

[78] T. H. Huxley 1861: 79.

for Owen would throw more than "a feather's weight into the scales" against himself, Huxley urged one of the foremost governing bodies of the scientific world to settle the hippocampus controversy as an issue of gentlemanly manners.[79]

As Rupke has emphasized, the criticism that Huxley began to level at Owen's comparative anatomy and paleontology (and at Cuvieran functionalism more generally) from the mid-1850s also served as a check to Owen's institutional power. In 1858, Huxley spearheaded explicit opposition to Owen's museum reforms in the form of a memorial to the chancellor of the exchequer, signed by "Zoologists and Botanists," including George Busk, William Carpenter, John Henslow, and (reluctantly) Darwin, in which Huxley advanced his own plan for a reorganization of Britain's natural history collections.[80] Referring to this museum memorial, which Huxley was helping to draft in October 1858, Carpenter wrote:

> I am not at all clear that the headship of a single man reporting direct to the Minister would be the best Government of the concern. Doubtless it works well at Kew; but who can find fault with the Hookers? Of course Owen would be the Autocrat of Zoology & Palaeontology; would it not be desirable to make him feel that he is responsible to a body of scientific men, who are competent to estimate and criticise his proceedings?[81]

In waging these campaigns, Huxley and others of the "young guard" may have been jealous of Owen's power, and, as Rupke has argued, attacked Owen's work and reputation to make names for themselves. But the criticisms of Owen by Huxley and other naturalists went deeper than designs to gain institutional power. Their criticisms were about the very structure of institutionalization that was proper to the study of the living world. The scientific research of those men who opposed Owen's museum plan, men such as Huxley, Carpenter, and Busk, had very little to do with either the shaping of the museum as a place of collection and display or the reconstructive work that had earned Owen his fame. Their work was situated in a different space, and indeed within a different science, one that was well established in Germany but had little institutional footing in Britain – namely, biology, which was a science that entailed laboratory practice.[82]

---

[79] Huxley to Sharpey, 13 November 1862, and 16 November 1862, Sharpey Correspondence, University of London, Senate House Library, MS Add. 227: 122, 124.

[80] Darwin to Huxley, 23 October 1858, in Burkhardt and Smith eds. 1985–2001, 7: 175–7.

[81] Carpenter to Huxley, 22 October 1858, HP: 12.94. The memorial is reprinted in Burkhardt and Smith, eds. 1985–2001, 7: 524–9.

[82] On the institutionalization of biology and laboratory practice in German universities, see Nyhart 1995. The extensive traffic in developmental theory between Germany and Britain has been explored in R. Richards 1992.

Huxley worked in a museum until 1870, and his first duties included the cataloguing of paleontological specimens. But his own reform efforts within the School of Mines were largely devoted to the removal of the institution from the museum site, to the installation of laboratory space, and to the use of the laboratory for the teaching of biology.[83] Almost from its beginning, Huxley's career was marked by his pleas for the preeminence of the laboratory over the museum as a space for scientific research and teaching. In one of his first public addresses, "On the Educational Value of the Natural History Sciences," delivered at St. Martin's Hall in 1854, Huxley said nothing whatsoever about museums either as research centers or as institutions for public instruction. Rather, his lecture introduced "Biology" as the science of individual life, "the experimental science par excellence," which, through physiological investigation, evinced the beauty and order of God's creation *in the laboratory*.[84] In a lecture that he gave in 1876, Huxley characterized the existing natural history museum, with its "happy hunting-ground of miles of glass cases," as a "splendid pile" that left the viewer with sore feet and a bad headache. Organized according to "the ideas of the bird-stuffer," it gave an impression of the animal kingdom as a "mighty maze without a plan."[85]

As Graeme Gooday has convincingly argued, Huxley and others appropriated the rational, recreational literature on microscopy, developed in the 1850s and 1860s by Charles Kingsley, Philip Gosse, Carpenter, and others, to promote the use of microscopes for teacher training and mass education, effectively transporting the pastoral virtues of the field to the enclosed and largely urban space of the institutional laboratory.[86] In the summer of 1871, Huxley and his team of demonstrators, which included Michael Foster, E. Ray Lankester, and W. T. Thistleton-Dyer, began a laboratory practicum for science teachers in a newly built facility at South Kensington.[87] Instruction commenced at 9 A.M. with a lecture by Huxley, and students passed the entire morning and afternoon in a rigorous course of dissection, microscopical observation, and drawing to scale, culminating in an examination that scrutinized their drawings and lecture notes for strict adherence to the opening discourse.[88] In

---

[83] Forgan and Gooday 1996.

[84] T. H. Huxley 1854c: 50.

[85] T. H. Huxley 1876: 284. See also his criticism of the Jermyn Street facilities given in evidence to the Royal Commission on Scientific Instruction, *Parliamentary Papers, 1872*, vol. 25, q. 296–7.

[86] Gooday 1991 and 1997. See also Desmond 2001.

[87] The importance of this course as a model for future teaching in the life sciences is discussed in Geison 1978: 130–47. The building of the South Kensington complex is examined in Forgan and Gooday 1994.

[88] See the account of the first six-week course at South Kensington in Lankester 1871: 362.

their pleas before liberal government officials for public support for the course, Huxley and his associates emphasized the virtues of the laboratory apparatus in fostering self-discipline and an appreciation for natural order, in contrast to the often arbitrary or tyrannical authority exercised by books and lectures.[89] Huxley's course, still based on a scale of natural historical types, did little to advance a theory of evolution – indeed, to do so would have been highly controversial and called attention to his own authority.[90] Instead, Huxley used the laboratory to replace traditional allegiances to texts, personalities, and social superiors. The laboratory was presented as a pure sphere, free from social influences like manners, in which one's relationship to nature was mediated only by impersonal and perfectly neutral instruments.[91]

Some historians have characterized the shift from a museum-based natural history to a laboratory-based biology as a fundamental transformation in nineteenth-century science.[92] By the 1850s and 1860s, museums had become important sites for the practice and teaching of science. In such spaces, which typically incorporated work rooms and large lecture halls, scientific research and instruction were structured around the museums' collections. The new pedagogical regime of the laboratory emerged, as Sophie Forgan has aptly described, in the shadow of the museum, and in competition with it.[93] It should not be surprising that this conflict was of central importance to the controversies between Huxley and Owen. Owen's success in museum negotiations had given him a monopoly over certain natural historical resources that no one could match. But more significant still, Owen's cultural role as a man of science for large sectors of the British public – namely, as the presiding figure in a temple of nature – was a role that could be filled only by someone who commanded enormous institutional resources, and Owen's efforts were directed toward ensuring his mastery of just these sorts of resources. Owen's position was not one that Huxley could hope to obtain, for it had been acquired largely through the reconstruction and display of rare, often massive, life forms. In the institutions of natural history within which Owen occupied positions, Huxley's kind of

---

[89] See, for example, *Parliamentary Papers, 1872*, vol. 25, Appendix xvii, pp. 54–5.

[90] Caron 1988: 250.

[91] The establishment of criteria of scientific objectivity and neutrality based upon a recourse to instruments is discussed in Daston and Galison 1992.

[92] See, for example, Benson 1991 and Pickstone 1994. On the demise of natural history and the rise of biology, see Allen 1978 and Caron 1988; however, see also Nyhart 1995, which argues for the transformation rather than the decline of natural history at the end of the nineteenth century. On the polemical importance of the laboratory as a model for scientific research more generally, see Jardine 1992.

[93] Forgan 1994.

science would always be subordinate. In size and grandeur, the building that housed Huxley's new laboratories at South Kensington was indeed overshadowed by the new British Museum of Natural History across the road, which opened in 1881 with Owen as superintendent. Owen ended his life at the head of a huge, expensive institution, his work displayed to thousands, supported by the leading figures in government, and attacked only by a handful of biologists who desired funding for laboratory facilities in universities and schools. Given that the history of natural history in Owen's hands is such an institutional success story, it is ironic that scholars have tried to repair his reputation in the history of science by presenting him as an evolutionary theorist and forerunner of Darwin.[94]

Why Darwin's *Bulldog?*

A year after the Museum of Natural History opened to the public, its grand entrance hall was graced with a memorial statue of Darwin (see Plate 4). Speaking at the museum in his capacity as president of the Royal Society, Huxley announced that the statue had been commissioned in conjunction with the establishment of a "Darwin Fund" for "the promotion of biological research." No better place for the statue could be found, he stated, than the museum's "great hall," where it could serve as a symbol, not of the "official sanction" of Darwin's views ("for science does not recognise such sanction"), but "of the ideal according to which [future students of Nature] must shape their lives."[95] Many of the museum trustees whom Huxley addressed at the unveiling were long-standing supporters of Owen. Whether the event should be seen as a Darwinian invasion of the sanctuary that Owen had labored for more than twenty years to build, or as an appropriation of Darwin by Owenites, is uncertain. It is clear, however, that Huxley tried to shape the meaning of Darwin's memorial in the cathedral of British natural history as a monument to scientific character rather than as a triumph of Darwinian doctrines. In similar maneuvers throughout his career, by which he transposed questions of scientific theory into questions of scientific conduct, Huxley made it impossible for men like Owen and Darwin to reconcile their differences. Huxley's new version of the man of science cut through older bonds of gentlemanly behavior and correspondence, fundamentally reshaping the social world of Victorian natural history even within the scientific community itself. But Huxley could not

[94] See Ruse 1979 and Rupke 1994: ch. 5. Compare also Desmond 1982: 72–4.
[95] T. H. Huxley 1885a: 251–2.

UNVEILING THE STATUE OF THE LATE CHARLES DARWIN IN THE NATURAL HISTORY MUSEUM, SOUTH KENSINGTON

Plate 4. Huxley delivering a speech at the unveiling of the Darwin statue, British Museum of Natural History
(*The Graphic*, 20 June 1885).

construct a new image for men of science simply by denigrating Owen. In opposition to the figure of Owen as the ambitious and corrupt autocrat, he needed a sympathetic figure of imposing purity and propriety. Thus his personal attacks on Owen proceeded alongside his enthusiastic tributes to Darwin. The elevation of Darwin as a heroic man of science and the denigration of Owen were part of the same process – each was a character witness in a trial staged by Huxley to weigh the merits of the man of science.

To a large degree, Huxley's image of Darwin as "the incorporated ideal of a man of science" was simply the positive to Owen's negative. Thus where Owen was a slavish follower, Darwin was "singularly original"; where Owen was given to airs and obfuscation, Darwin was "candid" and "clear," and where Owen had been diverted from the pursuit of knowledge to the pursuit of power, Darwin practiced "unswerving truthfulness and honesty." Huxley's portraits were not personality sketches; they were prescriptions for scientific manners and institutions. Huxley's characterizations implied that scientific creativity, discovery, and originality were destroyed by the structure of patronage. Through his rendition of Darwin as "outspoken," Huxley's own plain-speaking manners, which is to say, his aggressive and often personal attacks on men of established positions and reputations, were legitimized as a new code of scientific conduct, while the behavior proper to Owen's (and Darwin's) generation was called into question.

What made Darwin so exemplary for Huxley, and so effective a foil for Owen, was precisely what was most peculiar about Darwin, namely, his lack of institutional status and connections. A country squire, though one with his own elaborate network of correspondence, Darwin had little involvement with the institutional structure of patronage.[96] Nor did he monopolize institutional resources. Because he was not a new professional practitioner but was a gentleman of independent means, his science could appear quite free of any of the deleterious associations that came from using it to earn a living. Thus Darwin of Down was removed not only from the gentlemanly clubs, the politics, and the commercial pressures of the metropolis but also from the institutional fabric, increasingly interwoven with government and industry, of British science itself. Darwin's physical and social isolation facilitated his portrayal as a genius, for it was precisely this kind of environment – solitude – that, according to enduring cultural conventions, nurtured the intellect.[97] What was specifically Victorian about Darwin's genius was its place of seclusion – the home – its plain and simple manner, and its hard work.

[96] On Darwin's withdrawal from metropolitan society, see Rudwick 1982: 188–9.
[97] The importance of solitude in Western philosophy and the sciences is discussed in Shapin 1990.

Throughout the Victorian period, men of science featured the home as a place of work. In doing so, they had portraits made of themselves seated in their studies, usually at a desk, surrounded by the materials of learning – books, papers, and specimens – and occasionally accompanied (that is, assisted) by their wives. But while other men of science, like Owen, Huxley, or Tyndall, were also painted as popular lecturers or as public instructors, Darwin's portraits had the home as their exclusive setting (for example, his work-strewn study in the *Life and Letters*).[98] Darwin appeared in these portraits virtually as a disembodied thinker, withdrawn from the world, and at work in a domestic sanctum. Darwin's home was also a laboratory. His study had long served as a place for his microscopical work and dissection. After the publication of the *Origin*, much of his time was devoted to extensive botanical experiments, some of which he undertook at Huxley's suggestion.[99] Darwin's greenhouse, constructed in 1862 for his work on climbing plants, was featured in *Century Illustrated Monthly Magazine* in 1883, the year after his death.[100] At Down House, laboratory life proceeded with utmost purity, singular devotion, and a detachment from the politics of institutions like Huxley's, which were deeply embedded in such commercial and industrial programs as training engineers to exploit colonial resources. Thus while Huxley's campaign for institutional reform drew upon the Germanic model of a professional research laboratory, Darwin's laboratory practice manifested certain crucial Victorian values that could not be gleaned from the continental tradition. Located in the private sphere of domestic life, it could be free of the stigma attached to the pursuit of gain and advancement that characterized the public sphere.

In Huxley's particular rendition, the purity of purpose that Darwin obtained through his withdrawal from social circles and institutions also entailed a freedom from the codes of manners that had long regulated (and corrupted) gentlemanly associations. As a person of a "wonderfully genial, simple, and generous nature," Darwin escaped the political and material interests, and the veils of politeness, that impeded Owen from the pursuit of truth. Moreover, for Huxley, the taint of leisure and idleness clung to the older, gentlemanly ways that Darwin, with his unmitigated work discipline, had thrown off. Darwin's "genius," as manifested, for example, in his "tenacious industry" and "prodigious labours of original investigation," was in fact an elaborate Victorian invention;

---

[98] For portraits of men of science, see, for example, L. Huxley ed. 1900, Eve and Creasey 1945, and F. Darwin, ed. 1887. For a study of Darwin's portraits, see Browne 1998.

[99] On Darwin's botanical research as a response to Huxley's criticism of *Origin of Species*, see Burkhardt and Smith eds. 1985–2001, 10: Appendix VI.

[100] The image is reproduced as the frontispiece to Burkhardt and Smith eds. 1985–2001, vol. 11.

for eighteenth-century writers, genius had been an effortless capacity, usually existing in juxtaposition to hard work.[101]

As the figurehead of an entire, yet quite differentiated, scientific community, Darwin thus came to embody the twin qualities of genius and industry that Forbes had located in Murchison and the Geological Survey men. Huxley's biographical sketches of Darwin, like Tyndall's of Michael Faraday, portrayed their heroes as rare, exemplary individuals, representative of virtues characteristic of an entire community of scientific pracititioners.[102] In conjunction with such accounts, group portraits or sociologies like Francis Galton's *Hereditary Genius* and *English Men of Science: Their Nature and Nurture,* helped to shape images of the scientific community as a whole.[103] For Galton, Darwin's cousin, the potentially populist implications of the new scientific manners – the critique of gentlemanly propriety and the emphasis on individual merit, hard work, self-assertion, and self-attainment – were resolved by making "genius" an innate and inherited capacity (like the "blue blood" of nobility).[104] In Galton's work on hereditary genius, men of science became aristocrats of the mind. The powers that they possessed, the path into their elite world of science, were ultimately mysterious and unteachable.

Conclusion: Rag-and-Bone Men

Throughout the 1850s and 1860s, when Huxley's controversies with Owen were raging, there was no clear consensus among the community of practitioners about who was a proper gentleman and man of science. In 1862, following remarks by George Rolleston at the British Association in which he attacked Owen for ignoring the work of foreign anatomists on distinctive features of the human brain, an *Athenaeum* writer reported, "[Rolleston] felt there were things less excusable than vehemence; and that the law of ethics, and the love of truth, were things higher and better than the rules of etiquette or decorous reticence."[105] Yet, as has been discussed, many of Huxley's colleagues and friends had reservations about his own behavior. Charles Lyell admonished Huxley for his scientific conduct, urging him to remove certain paragraphs from his manuscript of *Man's Place in Nature* that were "not in good taste and

---

[101] Huxley 1882a. On the contrast between Victorian and earlier constructions of "genius," see Chapter 1.
[102] Tyndall 1868: 6, 180, 189.
[103] Galton 1869 and 1874.
[104] Galton 1874: 149–92, 232, 260.
[105] *Athenaeum,* 11 October 1862, p. 468.

will do no good."[106] On receiving a copy of the book, Hooker commented to Darwin on its evident coarseness, noting in particular a woodcut of a cannibal's "butcher shop" that Huxley had included as evidence of the savagery of which humans were capable. Hooker reported that the naturalist Hugh Falconer, "who has the most delicate and refined sense in such matters," was "disgusted" with Huxley's illustration "and would let no young Lady look at it."[107] Owen's own accounts portrayed Huxley as an ambitious power seeker who, by hiring a private coterie to stir up artificial emotions, behaved like a false politician. In an 1862 letter to Henry Acland, who was then head of the Medical Faculty at Oxford, Owen likened Huxley to the clever Athenian with "the itch of notoriety" who asked the oracle how to become a great man and was answered, "Slay one."[108] Responding to Tyndall's efforts in 1871 to bring about a rapprochement between the men, Owen wrote:

> Prof. Huxley disgraced the discussions by which scientific differences of opinion are rectified by imputing falsehood on a matter in which he differed from me. Until he retracts this imputation as publically as he made it I must continue to believe that, in making it, he was imputing his own (base and mendacious) nature.[109]

Accounts of the hippocampus controversy in the press and in fiction also centered on the question of gentlemanliness. Most of these favored Owen, while a few failed to distinguish between the men, mocking both. The *Lancet*, in 1862, accused Huxley of "falling back upon the usual arts of the controversialist, and importing passionate rhetoric to lead away the reader from the simple scientific questions at issue." It recommended that Huxley "try to imitate in these discussions the calm and philosophical tone of the man whom he assails."[110] Reporting on the Cambridge meeting of the British Association in 1862, the *London Quarterly Review* was surprised to find that, "in one of the two principal University towns in England," men of science comported themselves like the animals they studied: "Strange sight was this, that three or four most accomplished anatomists were contending against each other like so many gorillas."[111] In Charles Kingsley's *Water Babies* (1863), Huxley and Owen were satirized together in the character of Professor Ptthmllnsprts ("Put them all in spirits") who

---

[106] Lyell to Huxley, 9 August 1862, HP: 6.66.
[107] Hooker to Darwin, 26 February 1863. In Burkhardt and Smith eds. 1985–2001, 11: 179.
[108] Owen to Acland, 9 October 1862, cited in Rupke 1994: 295.
[109] Owen to Tyndall, 14 June 1871, cited in Rupke 1994: 295.
[110] *Lancet* (1862): 487.
[111] *London Quarterly Review* 19 (1863): 365.

had even got up once at the British Association, and declared that apes had hippopotamus majors in their brains just as men have. Which was a shocking thing to say; for, if it were so, what would become of the faith, hope and charity of immortal millions? You may think that there are other more important differences between you and an ape, such as being able to speak, and make machines, and know right from wrong, and say your prayers, and other little matters of that kind; but that is a child's fancy, my dear. Nothing is to be depended upon but the great hippopotamus test.[112]

Even more defamatory was a play published the same year by an author writing under the pseudonym George Pycroft, *A Sad Case, Recently Tried before the Lord Mayor, Owen versus Huxley*, in which the two combatants were caricatured as rag-and-bone men and arrested for a public disturbance. Appearing before the magistrate, Owen was called upon to prove his superiority to an ape by "the practice of kindliness, gentleness, forbearance, and humility," while Huxley was asked "whether it really was truth . . . he was fighting for [and] not more, or at least partly, for the purpose of exposing a weak spot in his rival."[113] As a *Punch* cartoon from 1865 showed, other practitioners delved into the layers of the earth's history, or probed the fundamental properties of matter, while Huxley and Owen wrestled only with each other, their simian objects of study looking on in amusement (see Plate 5). The spectacle of the two rivals continued to feature as a main attraction of meetings of the British Association for the Advancement of Science, even when neither man was scheduled to give a paper or chair a session.

Evidently, the question of manners was central, in the view of all these commentators, to the ability of men of science to represent the natural world truthfully. That the issue of manners no longer enters into scientific disputes at all should alert historians who persist in evaluating the character of Huxley, Owen, and other Victorian actors according to a code of scientific conduct that only came to prevail later. In retrospective accounts written toward the end of the century, the manners on display in these mid-century controversies would be completely transformed. In his *Autobiography*, Darwin recalled none of his working debts and cordial relations with Owen prior to 1860, but described him rather as someone who was always distant and peculiar, who was corrupted by jealousy, and who eventually became his enemy.[114] Contemporary journal accounts of their famous debate at Oxford in 1860 deemed Huxley "incourteous," and judged Wilberforce's speech a creditable performance.[115]

---

[112] Kingsley 1863: 156.

[113] [Anonymous] 1863, p. 4. For a survey of literary and press accounts of Huxley and Owen, see Blinderman 1970.

[114] Barlow ed. 1958: 104–5.

[115] *The Athenaeum*, 1705 (30 June 1860): 26, and 1706 (7 July 1860): 65; *Jackson's Oxford*

Plate 5. The prospect of Huxley wrestling Owen was still a main
attraction at the 1865 meeting of the British Association for the
Advancement of Science in Birmingham. Among the others featured
are John Tyndall, Roderick Murchison, Charles Lyell, and
Michael Faraday (*Punch*, 23 September 1865).

A complete inversion of these reports appeared in a reminiscence composed in 1899 by the liberal Anglican Frederick Farrar. Farrar claimed
that Huxley had been victorious on the basis of his "manners and good
breeding" and that the bishop "had forgotten to behave like a gentleman."[116] To many of his contemporaries in the 1860s, the new form of
self-presentation that Huxley offered as appropriate to men of science,
as manifesting their frank, open, and honest natures, still seemed unruly
and discreditable. By the end of the century, however, practitioners like
Owen, who had once seemed honest and polite, appeared disreputable
and ill-mannered.

Such transformations in scientific identity were also achieved in the
realm of institutions. By applying the laboratory model, which had
been successful in German universities, and which was also being extended to medical research in Britain, to the life sciences, Huxley worked
to diminish the importance of the museum as a site of ongoing research. In Huxley's writings and reform schemes, the museum became

Journal, 7 July 1860. For a survey and assessment of contemporary accounts of the
debate, see Lucas 1979.
[116] Farrar to Leonard Huxley, 12 July 1899, HP: 16.13.

preeminently a place of popular amusement and public instruction, suitable for the illustration of theoretical work performed elsewhere. At the same time, the heroic reconstruction of massive life forms that had been so crucial to Owen's institutional success was supplanted by the dramatic search for the microscopic principles of life. As objects of scientific inquiry and performance pieces, the moa and the megatherium gave way to protoplasm and the cell.[117] Several of Huxley's apprentices at the new science school in South Kensington went on to found their own laboratories and research centers in Britain – Michael Foster as professor of physiology at Cambridge, S. H. Vines as professor of botany at Cambridge, and Lankester as professor of zoology at Oxford. Darwin himself never proposed a science of biology to replace natural history. But as popularizers like Huxley suggested, Darwin's theory could reorganize much of comparative anatomy, morphology, and embryology into new schools of research. Indeed, Huxley's principal criticism of the theory of evolution by natural selction – that it had yet to be demonstrated experimentally – effectually subordinated historical and geographical fields of inquiry to the laboratory. Thus Huxley reoriented Darwin's theory, just as he redefined Darwin's image and reworked Darwin's relationship with Owen. In so doing, he shaped an institutional space and modes of conduct for a new kind of practitioner.

[117] See especially Huxley 1868a. On the role of Huxley and others in debates during the 1860s and 1870s over the origin of life, see Strick 2000.

3

Science as Culture

Perhaps the very silliest cant of the day, is the cant about culture.[1]
                                                      – Frederick Harrison, 1867

By presenting Owen as a person corrupted by power and self-interest, and by portraying the reclusive Darwin as the epitome of the pure and detached researcher, Huxley helped to shape the identity of the scientific practitioner as autonomous from society. But Huxley and the new generation of researchers he helped to train were in fact much more like Owen, employed in government institutions or on state-sponsored projects, embedded in institutional politics, implementing imperial policies. Perhaps even more than Owen, Huxley embraced the public role of science and sought to advance it. In addition to his duties at the School of Mines and Geological Survey, his teacher-training courses, and his public lectures, Huxley worked as a Science and Art Department examiner and as an inspector of fisheries; he eventually served on seven royal commissions, reporting to the state on matters of education reform, the fishing industry, and vivisection. In these posts, he employed his particular expertise in marine zoology and as a teacher of science. But he also developed for the man of science a broader agenda as cultural critic and commentator, competent to intervene in matters of general public interest or concern. Secured by scientific methods and procedures – acquired and honed in the laboratory – against any corrupting social interests or

[1] F. Harrison 1867: 610.

influences, the man of science, Huxley claimed, was the best of public
servants.

Yet the very process of promoting the man of science as socially autonomous and authoritative involved Huxley in extensive engagements
with other social groups. As the first chapter has already suggested,
scientific identity was not determined by elite practitioners alone. The
satirical accounts of the hippocampus controversy that appeared in periodicals and literary works during the 1860s raised serious questions
about the cultural authority and autonomy that issued from scientific
expertise. The kind of critical voice that Huxley sought for men of science
had also been claimed, and was in some measure held, by men of letters.
In the middle decades of the century, editors, publishers, and writers
could play a significant role in settling questions of scientific conduct
and institutionalization. Thus Huxley had to enter into a new set of relations with such figures to secure a position of broad cultural authority
for himself.

That Huxley is far better known today for his essays on politics, philosophy, and religion than for his vast body of technical papers and
monographs is perhaps a testament to the success of his campaign to promote science as culture. In their biographical accounts, his son Leonard
and his grandson Aldous, both literary men, emphasized Thomas's literary talents and accomplishments.[2] Subsequent scholars have noted
his linguistic abilities, his mastery of prose style, and the broad scope of
his knowledge and reading.[3] But Huxley's status in the world of learning and letters might be viewed as symptomatic of the growing power
and importance of science itself in the Victorian period.[4] It is perhaps
surprising, then, that a number of Huxley's contemporaries in literary
and philosophical circles regarded him as a narrow specialist, lacking
the breadth required of the general critic or educational reformer. Nor
was it the case that either "science" or literary "culture" were widely
accepted as authoritative in matters of politics, public welfare, or education. This chapter will examine Huxley's efforts to establish himself as a
man of letters while maintaining his role as man of science, and the implications of this program for his definition of scientific identity. A look
at Huxley's early activities as a science writer and reviewer for the periodical press will reveal how difficult it could be to translate scientific
knowledge and expertise into authority in the literary domain, expe-
cially where the criteria of literary authority were uncertain. A study of

---

[2] L. Huxley ed. 1900, 1: 297–8, 2: 25–6, 85, 419–20; A. Huxley 1932.

[3] See, for example, Bibby 1959: 16, 20–1, 42, and Barr ed. 1997b: xxi.

[4] Among the accounts that argue for, or assume, the increasingly normative role of science
in Victorian culture, see Cannon 1978: 1–28, Paradis and Postlewait, eds. 1981: ix–xiii,
Heyck 1982, and Levine ed. 1987: 8–12.

Huxley's later engagements with other men of letters, Matthew Arnold in particular, will show how new definitions of "culture" as a product of scientific and literary elites provided a common ground for men of science and letters as authorities in the sphere of education.

Science Writing and the Periodical Press

In practical terms, the Victorian man of letters was very often a writer for the periodical press. From the early decades of the nineteenth century, journals like the *Edinburgh Review* and the *Quarterly Review* had cast themselves as educators of public opinion. By identifying themselves as the exclusive producers of "culture,"[5] writers such as John Stuart Mill, Samuel Coleridge, and Thomas Carlyle had defined the role of the "clerisy" or "man of letters" as that of an authority and educator of the commercial and industrial classes. But the status of such elite definitions of culture was by no means certain in the periodical domain, particularly by mid-century, when review writing had become both a major profession in its own right and an important source of income for aspiring novelists, philosophers, and men of science.[6] Huxley's first mentor, Edward Forbes, had supplemented his small salary by contributing regular reviews to a variety of journals such as the *Literary Gazette*.[7] Huxley, before he gained institutional footholds at the Museum of Practical Geology and School of Mines, had supported himself partly through journalistic activity and translation. Late in 1853, he was approached by the publisher John Chapman to contribute a regular column on science for the *Westminster Review*.[8]

Huxley's position on the *Westminster Review* had been facilitated by his contacts with a circle of London writers who often met at Chapman's home. The group included Marian Evans (soon to adopt the literary guise of George Eliot); the playwright, actor, and journalist George Lewes; the phrenologists Charles Bray and George Combe, and the young philosopher Herbert Spencer.[9] Members of this circle had similar backgrounds. Having acquired their educations outside the universities of Oxford and Cambridge, they found that, as a result, many avenues into elite circles of learning were closed to them. They shared a desire to reform the culture one acquired with an Oxbridge degree, a culture

---

[5] For discussions of the Victorian "clerisy," see Raymond Williams 1958, Kent 1969, and Heyck 1982.

[6] On the profession of writing in the Victorian period, see Gross 1969 and Cross 1985.

[7] See Forbes 1855.

[8] Chapman, letters to Huxley, 23 and 26 October 1853, HP: 12.169–70.

[9] On the *Westminster* circle, see Ashton 1996: 135–63 and Haight 1940: 28–40.

centered primarily on familiarity with the classical canon. Many of the *Westminster* contributors had shown a keen interest in the writings of the French philosopher Auguste Comte and in his platform for social improvement through a progressive elaboration of the sciences.[10] Thus Chapman's journal became a medium for extending the boundaries of literature and science; in the *Review*, specialist knowledge and continental fiction were digested for a nonspecialist audience and incorporated in broad programs for human understanding and reform.

Writers for the *Westminster Review* were largely dependent on the income they derived from their articles. This was not true for writers with some of the other literary periodicals, such as the *Quarterly Review* or the *Saturday Review*, whose contributors either were largely Oxbridge-based or held positions in one of the established learned professions, such as the bar or the Church. As a profession, periodical writing was a risky venture; practitioners without institutional status or independent means could be stigmatized as "popular" writers, crass journalists, or traders in knowledge and literary fashion.[11] Unable to claim the distinction that derived from attending established institutions of learning or entering revered professions, and utterly reliant therefore on the dictates of the literary market, *Westminster* writers nonetheless sought to present themselves as independent authorities, distinct from writers who, according to Lewes, appealed only to a "large and listless public which must be amused; and ... is far from critical."[12] Toward this end, they incorporated scientific knowledge and practice into their literary activities, thereby redefining the criteria for serious criticism and original fiction.

In his 1852 essay "The Philosophy of Style," Spencer derived a series of rhetorical principles based on the effect of various forms of composition on an audience.[13] From the 1840s, Lewes had been outlining a similar method of criticism based on understanding the responses of the senses and feelings to works of fiction.[14] In so doing, he drew on his own extensive reading of associationist psychology and continental physiology. In a series of essays written in the late 1850s and 1860s, Lewes developed this approach: a reviewer of literature needed to apprehend the scientific principles that made a piece of writing work – to be a kind of

---

[10] On Comtism in Victorian Britain, see Wright 1986 and Dale 1989. On the later, Oxford-based movement, see Kent 1978.

[11] The problematic relationship between scientific and commercial writing is carefully examined in J. Secord 2000: chs. 13–14.

[12] Lewes 1858: 500.

[13] Spencer 1852b.

[14] See, for example, Lewes 1842b: 481. On physiological models of reading, see Johns 1996: 138–64 and Jones 1996: 75–87.

anatomist and physiologist of style.[15] Scientific materials and methods were used to similar effect by Eliot. In anonymous review articles for the *Westminster*, she began to advance a realist approach to fiction, based on "direct observation" and a profound knowledge of human nature, drawn chiefly from social science, psychology, and physiology.[16] She applied these principles in distinguishing serious fiction from novels that catered to current sentiments and unexamined conventions.

Huxley shared the commitment of these writers to a model of culture divided into elite and popular. It was precisely during these decades that "popular science" was emerging as a category of writing, with complex and contentious connections to expert scientific practice. At the time of his appointment to the *Westminster Review*, despite already possessing a reputation as an expert on the comparative anatomy of marine life, Huxley was still struggling to gain a position that would assure this reputation. He used review writing to carve out a role for the man of science outside the restricted sphere of specialist journals, societies, and laboratories. The periodical, however, was a volatile domain in which the authority of gentlemanly specialists and institution-based experts was not clearly established. In this literary world, distant from the traditional sites and sources of scientific authority, how were the claims of specialist practitioners to retain a different status from those of other writers?

In his science columns for the *Westminster*, Huxley drew sharp distinctions between works of genius, which produced new knowledge, works of popularization, which disseminated knowledge, and works of popular delusion or superstition, which debased knowledge. Favorable reviews were written of works whose "charming wonderful stories" were calculated to instill the importance of natural historical knowledge in children, or of guidebooks for the seaside collector, such as Philip Henry Gosse's *Manual of Marine Zoology for the British Isles*. Such works were to be read as "introductions," rather than "rivals," to the elaborate monographs of George Busk, Edward Forbes, Thomas Bell, and other experts on British marine zoology.[17] By contrast, works advocating mesmerism, spiritualism, and clairvoyance were accused of spreading a popular delirium for which there was but one remedy: "early education...in the methods of the natural sciences and the inculcation of inductive habits of mind."[18] Such categories of authorship served to divide the readership for science into an elite community of practitioners on the

---

[15] Lewes 1858 and 1865.

[16] Eliot 1856a, 1856b. On Eliot's use of natural historical and experimental models for fiction, see Shuttleworth 1984.

[17] T. H. Huxley 1854–7, 61: 263.

[18] T. H. Huxley 1854–7, 65: 605.

one hand and an uninstructed public on the other. Yet Huxley's reviews also worked to create a new audience that lay between the extremes of expertise and ignorance. It was a readership whose rudimentary and empirical knowledge, and whose "amateurish" participation, could affirm the authority of an elite community of specialists.[19] In order for readers to recognize such authority, however, they must be able to differentiate the popular or professional writer, whose goal was amusement and profit and whose mental powers were derivative and superficial, from the true practitioner, who was moved by genius and the pursuit of truth.

By commenting extensively on works written for a nonspecialist audience by authors who were ambiguously situated in the metropolitan scientific world, Huxley drew certain boundaries between the scientific elite and the "popular" and tried to position the man of science prominently in a rapidly expanding print sphere. In establishing criteria for scientific authority in the periodical domain, however, Huxley challenged the position of other members of the *Westminster* circle, whose criticism rested partly on their ability to appropriate and implement scientific materials through reading. By stigmatizing as "popular" those works that relied on the researches of others, and by discrediting works that routinely cited other scientific authorities in order to secure their own, Huxley was drawing a sharp distinction between scientific practitioners and ordinary reviewers. Huxley's very first article for the journal attacked a recent book by Lewes for displaying an ignorance of science that stemmed from the author's exclusively literary life. In this book, *Comte's Philosophy of the Sciences*, a collection of articles that had first appeared in another periodical, the *Leader*, Lewes had set out to explicate Comte's work of the 1830s and to bring it up to date with the latest facts and theories from chemistry and physiology.[20] Huxley picked away at Lewes's science, alleging errors such as the insertion of "sulphuric" for "sulphurous acid" and criticizing the favorable presentation of developmentalism, to which of course Huxley was opposed at the time.

> We are taking advantage of no accidental mistakes, although those already cited would suffice to show, if demonstration were needed, how impossible it is for even so acute a thinker as Mr. Lewes to succeed in scientific speculations, without the discipline and knowledge which result from being a worker also.[21]

---

[19] On the importance of participatory, as opposed to exclusionist, models of popular science, see especially A. Secord 2002.

[20] Lewes 1853.

[21] T. H. Huxley 1854–7, 61: 255.

Although Huxley commended Lewes's forceful and acute expression, considerations of style were outweighed in his assessment of the book as a work of science. According to Huxley, Lewes had committed errors of fact that no reading or writing, however thorough and careful, could correct. To be up-to-date in science, Lewes needed a specialist's training and experience in the field and laboratory. In effect, Huxley was claiming that the criteria that Lewes, Eliot, and others used to distinguish sound literature were inadequate for science.

Huxley's criticism turned as much on the medium through which knowledge was communicated as on its mode of acquisition. Original scientific work was typically presented in specialist journals that were available to a small circle of subscribers. It was read before learned societies, or published in lengthy, expensive volumes. The periodical press was not the proper forum for making scientific truth public. Rather, the periodical was a vehicle for a different kind of science writing, tailored for a different audience. The truth expressed in learned society monographs, published at a loss, and intended for the perusal of a small and select audience could gain substantial currency through such a medium. As an aspiring man of science, Huxley thus depended on periodical publication as a way of validating his laboratory expertise as cultural enterprise just as much as, he was claiming, Lewes needed personal laboratory experience to validate his public statements on science. For Huxley, it was of paramount importance not only to distinguish between the popular expositor and the scientific expert but to control the grounds on which this distinction was based.

In her capacity as assistant editor for the *Westminster*, Eliot read the proof copy of Huxley's review of Lewes and urged Chapman to suppress such a "purely contemptuous notice."[22] Even before seeing the article, she had written several letters to George Combe expressing her reservations about Huxley as a reviewer:

> Mr. Huxley's is not the organization for a critic, but it is difficult to find a man who combines special scientific knowledge with that well-balanced development of the moral and intellectual faculties, which is essential to a profound and fair appreciation of other men's works.[23]

Lewes himself responded to Huxley in a column in the *Leader*, objecting to being unfairly treated as a "bookman and not even a respectable bookman" and challenging Huxley's attempt to locate the boundaries of scientific expertise outside the world of literature:

[22] Eliot to Chapman, 17 December 1853, in Haight ed. 1954–78, 2: 132.
[23] Eliot to Combe, 16 December 1853, in Haight ed. 1954–78, 8: 91. See also Eliot to Combe, 23 November 1853, in Haight ed. 1954–78, 8: 89.

Appearing in the pages where it is well known I am also a writer... this attack will have more than usual significance; and being founded on the natural but false assumption that, because Literature is my profession, therefore in Science I can only have "book knowledge," it will fall in with the all but universal tendency of not allowing any man to be heard on more than one subject.[24]

The enterprise of the *Westminster* circle depended crucially on the authority of reviewers to span specialist domains, synthesize their contents, and comment on their worth. The realist fiction and literary criticism of Lewes and Eliot, as well as the philosophic essays of Spencer, Combe, and others, rested on extensive programs of reading and on observations of everyday life. A review writer required a scope of knowledge and sympathy that were unobtainable in the confined space of the laboratory, or as Eliot put it, a "moral and intellectual breadth" not "vitiated by a foregone conclusion or by a professional point of view."[25] From this perspective, Huxley's scientific expertise was merely that of a narrow specialist selfishly defending his terrain against others. Lacking the sympathy that was born of broad learning, Huxley was unqualified to comment on the work of those outside his particular field and thus unfit for the office of critic.

Huxley tried to vitiate such criticism by asserting the authority of the laboratory and field experience within the sphere of the periodical. If heeded, such assertions would have consigned Lewes, Eliot, and most other writers for the *Westminster Review* (and other periodicals) to the realm of popular writers or popularizers on Huxley's terms. This, of course, is precisely what *Westminster* authors were striving against, in large part through an appropriation of scientific methods and knowledge. Following Huxley's criticism, Lewes acquired a microscope and embarked on a systematic study of marine life, which resulted in a series of articles published in *Blackwood's Edinburgh Magazine* in 1856 and 1857.[26] Through affecting, first-person narratives, he gave readers step-by-step instruction in the use of scalpel and microscope, in conjunction with book knowledge. He also intervened in specialist debates on comparative anatomy and physiology and claimed original discoveries on the nervous system of mollusks. His literary rendering of scientific practice extended a bridge across the elite-popular divide, as Huxley had drawn it, making the tools of expert knowledge accessible outside of the institutional structures that specialist practitioners controlled. It also offered possibilities for the literary periodical

[24] *Leader* 5 (14 January 1854): 40.
[25] Eliot 1856a: 56.
[26] Lewes 1856 and 1857.

as an alternative forum to the specialist journal and learned society for *farebrother* scientific communication.

Lewes's work, republished in book form as *Sea-side Studies*, is but one example of a vast literature on the sciences that was written by persons outside or on the margins of the elite and increasingly institution-based scientific community. Though clearly operating within the newly designated genre of popular science, authors such as Margaret Gatty, Arabella Buckley, Richard Proctor, Frank Buckland, and Charles Kingsley produced volumes that rivaled or exceeded in influence those of Huxley, Tyndall, and other elites and that presented alternative views of nature and scientific participation.[27] It proved extremely difficult, even for recognized experts, to control the meaning of science and to police the boundaries of expert and popular, in the realm of literature. Ultimately, the identity of men of science as cultural authorities, independent from the demands of the literary market, would be secured through affiliation with elite institutions of learning and their classical curriculums – precisely those traditional structures to which the *Westminster* circle had been opposed.

Literature and Liberal Education

Huxley's brief sojourn as a review writer ended in 1856, as his institutional posts accumulated and his teaching responsibilities grew apace. Many of the concerns he had expressed in his science columns were now channeled into educational campaigns. In the domain of education, however, Huxley faced a rather different obstacle to that of narrow specialism. He had to contend with the view, widely espoused by legislators and major patrons of learning in the period, that science (with the exception of mathematics) was essentially a practical enterprise, valuable for industry, agriculture, and the material advantages of life but inessential or even detrimental to a "liberal" education of the intellect and character. At mid-century, the leading model of liberal education was still classical. The curriculum at elite institutions such as Oxford and Cambridge, and at the prestigious Anglican schools such as Eton and Harrow, was devoted to the Greek and Latin languages and, to a lesser degree, to mathematics. Oxford and Cambridge required proficiency in Greek and Latin for admission.[28] Most public schools were compelled by

[27] On popular science writing in the Victorian period, see Sheets-Pyenson 1988, Myers 1989, McRae ed. 1993, and Lightman 1997.

[28] On the classical model of liberal education, see Rothblatt 1976, Turner 1981, and Heyck 1982. On Oxford and Cambridge entry requirements, see Turner 1981: 5.

their charters to teach only classical subjects.[29] Arguments in favor of the traditional curriculum emphasized that the classical canon was the best source of mental discipline and moral values for England's future leaders. Even reformers like William Whewell, who were concerned with the status of the inductive sciences, resisted the incorporation of such "progressive" elements into the curriculum and defended the "permanent" subjects of classics and mathematics.[30] The classical model also prevailed at a great many other, less expensive and less renowned institutions. The University of London, which had been established as a secular alternative to Anglican Oxbridge and which offered a wider range of courses, including, for example, modern languages, still required Latin for admission.[31]

Huxley's own institution, by contrast, was oriented around a practical model of education that had prevailed in many industrial centers in England since the early Victorian period. The "School of Mines" (the actual name changed many times over the century) had been established in 1851 under the supervision of the Board of Trade. It was relocated administratively under the Science and Art Department when the latter was created in 1853.[32] The School of Mines, which offered an associate degree after two years of course work, was designed chiefly to provide training in the sciences for students wishing to enter technical professions such as industrial chemistry, engineering, and mining. It offered the kind of courses that had been available previously only at some Dissenting academies, private institutes, and hospital schools, or at universities abroad. As defined by the Committee of Council on Education, the institution occupied the uppermost echelon of a system of technical education for the industrial classes.[33] As a state-sponsored school of higher education in the sciences, however, it differed considerably from the Mechanics' Institutes and other "useful knowledge" institutions.[34] Its teaching staff were supported not by fees, but by regular salaries, and were given titles comparable to lecturers and professors at other universities.

Although many of the courses and professorships at the School of Mines were in "applied" sciences such as metallurgy, Huxley tried to

---

[29] On the classical curriculum of public schools, see Roach 1986: 238–9 and Shrosbree 1988. On public school charters, see Altick 1957: 178.

[30] Yeo 1993: 209–24.

[31] The University of London was purely an examining and degree-granting institution until 1900; its matriculation requirements and examination subjects are given in Harte 1986.

[32] See Bibby 1959: 124. For a the history of the School of Mines, see [Anonymous] 1945.

[33] See, for example, Eighteenth Report of the Science and Art Department of the Committee of Council on Education. *Parliamentary Papers, 1871*, vol. 24.

[34] On Mechanics' Institutes, see Inkster 1976 and Shapin and Barnes 1977.

orient the curriculum toward general science education, arguing for the dependence of practical pursuits on a knowledge of scientific princi- ples. Throughout his tenure at the institution, he negotiated with others within the school and the Science and Art Department, as well as with a variety of corporate patrons like the Livery Companies of London, over the material benefits of general science education. Natural knowledge, Huxley argued, had produced innumerable improvements, such as the steam pump and the spinning jenny, crucial for the production of wealth. What business had a manufacturer without a working knowledge of his own engines or of the nature of the raw products he employed?[35] Yet despite this concern with the material benefits of science, important to his reform efforts within the School of Mines, Huxley consistently subordinated the practical value of science to its moral and intellectual value. Commenting in 1864 on a parliamentary debate over the introduction of science to public school curricula, Huxley criticized politicians who naively questioned the value of science for mental training, and who patronized science only for its "froth and scum – its so-called practical results," while "scorning its essence and the foundation of its human worth."[36]

From the beginning of his career at the School of Mines, Huxley engaged in a broad campaign for the introduction of the sciences into English schools and universities. This platform involved the promotion of science as a part of "liberal" rather than "technical" education. By the mid-1850s, it was virtually a standing argument among liberal reformers that the classical curriculum was too restricted and that it was being taught in a manner that was too mechanical, by rote learning of examples rather than by training the mind to think independently. The Clarendon Report, issued in 1864 by a royal commission on public schools, stated that, although the classics had had "a powerful effect in moulding and animating the statesmanship and political life of England," they were now studied "only superficially" and gave results that compared unfavorably with the progress of science.[37] Science, the report continued,

> quickens and cultivates directly the faculty of observation ... the power of accurate and rapid generalization, and the mental habit of method and arrangement; it accustoms young persons to trace the sequence of cause and effect ... and it is perhaps the best corrective for that indolence which is the vice of half-awakened minds, and which shrinks from any exertion that is not, like an effort of memory, merely mechanical.[38]

---

[35] T. H. Huxley 1866: 29. See also Bibby 1959: 114–15.
[36] T. H. Huxley 1864.
[37] *Parliamentary Papers, 1864*, vol. 20, pp. 28–30.
[38] *Parliamentary Papers, 1864*, vol. 20, pp. 32–3.

Huxley followed this line of argument, and extended it considerably. He repeatedly lectured schoolmasters, Royal Institution audiences, and government officials on the value of science in forming the intellect and character. Science, he claimed, conferred a respect for observation and experiment rather than for authority; it taught the value of evidence and instilled a firm belief in immutable moral and physical laws. Science was also a "moral discipline," its success depending on the "courage, patience, and self-denial," rather than the mere "talent," of the practitioner.[39] Addressing an audience at the Science Museum at South Kensington in 1860, he remarked that an exclusively classical education, appropriate perhaps for the fourth century, was sorely out of touch with modern civilization, such that "even the mere man of letters, who affects to ignore and despise science, is unconsciously impregnated with her spirit, and indebted for his best products to her methods."[40] Appearing as a witness before the House of Lords Select Committee on Public Schools in 1865, Huxley declared that science was "an indispensable means of acquiring knowledge"; it imparted a "mental discipline" that was "not to be obtained otherwise."[41]

Throughout these shifts in emphasis and orientation, one thing remained constant in Huxley's characterizations of science – its juxtaposition to literature. As he had done in his efforts to establish credibility as a science writer for the periodical press, Huxley repeatedly stigmatized literary practices in the course of defining and dignifying the role of scientific subjects and methods in education. What he encompassed within "literature" in this context, however, was quite different. In periodical writing, the "literary" had denoted an exclusive devotion to books and matters of style, an acceptance of second-hand truths, and a tendency to promulgate popular delusions. But in the context of educational debates, Huxley identified literature with the existing classical curriculum of the universities and public schools. By extending the boundaries of literature to the highly formalized, rigorously structured study of ancient languages (and, implicitly, of mathematics), Huxley was able to deploy the same tactic that he had used in his science articles for the *Westminster Review*, namely, the juxtaposition of errors promulgated in books with truths gleaned through observation and experiment. If book-learning was delusive and dangerous in periodicals, enshrined in institutions of higher education it was positively tyrannical.

In 1866, Huxley was appointed to a committee of the British Association for the Advancement of Science to prepare a statement on the

---

[39] T. H. Huxley 1856b. See also T. H. Huxley 1854c: 59 and 1861: 226.
[40] "On the Study of Zoology," in Barr ed. 1997b: 17.
[41] *Parliamentary Papers, 1865*, vol. 10, pp. 306–11.

best means of promoting scientific education.[42] The committee's report, which was delivered in the following year to the president of the Committee of Council on Education, presented science as part of a reformed classical curriculum designed to elicit the highest moral character. It called for an equal representation of science in university examinations, honors, and fellowships and a minimum of three hours of science per week in all schools.[43] To justify the introduction of the sciences, the authors of the report juxtaposed a "mere literary acquaintance with facts" to scientific education, which imparted the "best discipline" in "accuracy of thought and language."[44] Testifying before the parliamentary Select Committee on Scientific Instruction in 1868, Huxley went still further, asserting that, "at present the universities make literature and grammar the basis of education, and they actually plume themselves upon their liberality when they stick a few bits of science on the outside of the fabric ... The thing you really have to do is ... to invert the whole edifice, and to make the foundation science, and literature the superstructure and final covering."[45] In an address delivered a year later, Huxley depicted the classical tradition as a "thrawldom of words," thereby facilitating the portrayal of science as the embodiment of true liberal principles – freedom of thought, anchored in disciplined observation.[46] Without continuous reference to the "world of facts," words (and numbers), the men who cultivated them, and the books and languages that contained them could become impediments, rather than aids, to learning.

What then was the proper place of literature in a reformed classical curriculum? In his 1868 address "On a Liberal Education," Huxley portrayed literature as a "source of pleasure without alloy" and a "serene resting place for worn human nature."[47] In casting literature as a consoling complement to arduous work, a place of rest and restoration from dreary struggle, and a source of inspiration for tasks ahead, Huxley was drawing on an extensive Victorian discourse that feminized and domesticated fiction.[48] In 1852, for example, Spencer had contrasted poetry with all things "active" and "useful." It was not productive, but "ornamental," "decorative," and "picturesque." Rather than work in the world, it brought "relaxation," "enjoyment," and "amusement in leisure hours."[49] Similarly, in his *Autobiography*, John Stuart Mill concluded that

---

[42] See Bibby 1959: 166–7.
[43] *Parliamentary Papers, 1867–8,* vol. 54, pp. 4–7.
[44] *Parliamentary Papers, 1867–8,* vol. 54, pp. 3, 5.
[45] *Parliamentary Papers, 1867–8,* vol. 15, p. 402.
[46] T. H. Huxley 1869: 129–30.
[47] T. H. Huxley 1868a: 96.
[48] On the feminizing and domesticating of fiction in England, see Poovey 1984 and 1988.
[49] Spencer 1852a: 371–3.

literature had cured him of his depression because it furnished "internal culture" to balance "the ordering of outward circumstances"; it nourished the "passive susceptibilities" to counter the "active capacities"; it provided a "cultivation of the feelings" to match "the training of the human being for speculation and for action." In effect, Mill accommodated literature by consigning it to private life.[50] By identifying fiction with leisure and repose, such usage divorced literature from sober, serious work, from social leadership, and from the foundations of learning. Thus in the future, Huxley asserted, literature would have to take its proper place as the "foliage and fruit" of education, while science supplied the "sound nutriment."[51]

By employing a scheme of separate spheres, analogous to the gendered dichotomies that defined his own relations of scientific work and home, and of husband and wife, Huxley tried to position science within the established tradition of liberal education. Through organic metaphors such as the tree of knowledge, with its scientific roots and literary leaves, Huxley depicted science and literature as complementary constituents of classical culture. Clearly there were issues of rank and authority to be negotiated within this model of liberal culture. The implied subordination of literary superstructure to scientific foundations was at one level a radical reversal of existing institutional hierarchies. Yet, at another level, Huxley's claim was not that science was superior to, or independent of, literature, but that each was dependent on the other. If the fruit did not nourish the tree, neither did the roots bear fruit. Likewise, at an institutional level, the measures that Huxley and other scientific reformers proposed were designed not to replace traditional liberal subjects, but to complement them. Huxley's rhetoric and policy of mutual support was not lost upon legislators and educators, many of whom were concerned about defending the established classical curriculum against encroachments of a practical or technical, or, indeed, of a secular, sort.

Huxley's educational campaigns have often been interpreted as democratic and utilitarian efforts to reform the elitist and outworn classical canon. His program of scientific education, much like his science popularization, has thus been aligned with the interests of the industrial and commercial classes against the entrenched privileges of the Anglican establishment. Yet it is clear from his participation in the debates over the reform of universities and schools, and from his efforts within the School of Mines itself, that he worked hard to detach science from associations of a narrowly practical character. His criticisms

---

[50] J. S. Mill 1961: 100–3.
[51] T. H. Huxley 1869: 115–17, 127.

of England's elite institutions did not turn on their failure to prepare students for careers in business or manufacturing, but on their reluctance to accord a place for science *within* the classical curriculum.

That the sciences should assume an important place in the broad Anglican culture of the universities and public schools was also a view promoted within these elite institutions. The liberal Anglican cleric Frederick Farrar, headmaster of Harrow, wrote a series of essays advocating the sciences as an effective means of intellectual training. In 1867 he served with Huxley and Tyndall on the British Association for the Advancement of Science's committee on scientific instruction.[52] Huxley's views on science education made him an ally of those advocating a more liberal and progressive Anglican establishment. Huxley cultivated ties with other leading Anglican reformers besides Farrar, such as Arthur Stanley, Benjamin Jowett, and Charles Kingsley. Huxley's relationships with liberal Anglicans are discussed in the next chapter for their bearing on issues of science and religion. It is important to note here, however, that ties with members of elite, classically oriented institutions were an effective means of contesting the practical image of science embodied in new institutions like the School of Mines. Such connections are exemplified in the close friendship and working alliance that developed between Huxley and Matthew Arnold in the 1860s and 1870s.

Friends and Enemies of Culture

Huxley met Matthew Arnold at the Athenaeum Club in 1867. Though his scientific friends Darwin and Hooker had been unwilling to nominate him (see Chapter 2), Huxley was eventually elected to the club in 1858 under a special provision that allowed a limited number of persons to be chosen by a committee rather than by the general membership.[53] Huxley's scientific publications were then only of a highly specialized and technical sort, but his Royal Society fellowship and royal medal were strong grounds for Roderick Murchison, who was a trustee of the Athenaeum as well as Huxley's superior at the Museum of Practical Geology, to bring Huxley forward as a candidate. The Athenaeum had had strong connections with science from its founding in 1824 by a group that included the president of the Royal Society, Humphry Davy. The group's aim had been to bring together the most distinguished men of letters, science, and the fine arts with those members of the nobility, gentry, and learned professions who were particularly known for

---

[52] See the Committee's report in *Parliamentary Papers, 1867–8*, vol. 54, pp. 4–7. On Huxley's ties with Farrar, see Bibby 1959: 166–7.

[53] Cowell 1975: 165.

their patronage of learning. With its palatial setting, its array of servants on call at all hours, its provision for billiards, brandy, and beef, the Athenaeum was like many other exclusive London clubs. But it was marked out as a place of privilege for gentlemen of culture by the predominance of its libraries and by its relatively modest fee of six guineas a year. The building itself, with its statue of Athena poised beneath a frieze of Parthenaic figures, and its Roman-Doric portico supported by six fluted columns, was designed for the heirs of classical culture.[54] The club offered some comforts of wealth and privileges of birth to men who were distinguished primarily by their learning. It also provided a place for persons otherwise widely separated by background and profession to associate.

Although Huxley and Arnold were both sons of schoolmasters, their education and upbringing had been quite different. Huxley had attended a lesser public school in Ealing, where his father taught mathematics. Although offering a traditional classical education, the Ealing school was evidently in a poor state, for Huxley had to leave after only two years when his father was unable to earn a living from the declining fees. Such formal education as Huxley had thereafter was entirely of a specialist sort in medicine and the sciences. He acquired a reading knowledge of Latin, some Greek, and other European languages, but he lacked the facility in use of the classics that might come with a traditional higher education. By contrast, Arnold had attended Rugby, one of England's most elite public schools, where his father had earned a reputation as a leading educational reformer and classical scholar. Having distinguished himself by winning prizes in Latin verse at school, Arnold went on to Oriel College, Oxford, where he obtained a two-year fellowship and more literary prizes. After leaving Oxford in 1847, Arnold published several volumes of poetry and supported himself through educational work. He served for four years as a private secretary to the marquis of Lansdowne, the head of the Committee of Council on Education. In 1852 he was appointed an inspector of denominational schools, that is, of those schools endowed by various Dissenting bodies.[55]

Through their divergent paths, Huxley and Arnold both found careers in education in the employ of the state. Like the School of Mines, the inspectorate in which Arnold served was a product of government intervention in a period when education was still dominated by voluntarism and private corporations. The inspectorate had been formed in conjunction with a series of measures, including the creation of an

[54] Cowell 1975: 8–18, 32.
[55] Honan 1981: 247, 263, 293–7, 341. For a detailed account of Arnold's career in education, see Connell 1950.

education council, designed to formalize the role of the government in education. Although the majority of educational institutions remained in private hands, the financial support that they received from the state had increased substantially over the first half of the nineteenth century. The middle decades were marked by protracted disputes over government regulation and expenditure in education.[56] As a means of making people into rational, self-reliant individuals, education had come to occupy an important place among liberal alternatives to state intervention and expenditure in other areas, such as poor relief and prisons. Likewise, in debates leading up to the Reform Bill of 1867, liberals campaigned to broaden the base of the educated public in conjunction with the expansion of the franchise.[57] Arnold was greatly in favor of an expanded inspectorate charged with regulating, and indeed elevating, educational standards. In his school reports, he pointed out that although many pupils possessed much "information," such as the facts of geography and history, they lacked "culture and intelligence." He recommended a study of "classic" English authors to "humanize" the pupils and thereby bring them into "intellectual sympathy with the educated of the upper classes."[58]

When Huxley met him in 1867, Arnold was just approaching the end of a ten-year tenure as professor of poetry at Oxford. Arnold had used the Oxford chair as a platform for an ambitious program of cultural reform. His 1860–1 lectures on Homer were widely reviewed and spread his reputation as a scholar and critic.[59] In subsequent lectures, on the function of literary criticism, he outlined the role of the scholar in ministering to the needs of a modern nation, independent of religious sects, party politics, or the interests of any profession or class.[60] In particular, he promoted men of letters as agents for reforming English society by refining and improving its sensibilities. He explicitly distanced his own cultural program from the platform of "middle-class liberalism," which he defined as the machinery of covenants and reform bills, of local self-government and free trade, all propelled by a worship of religious dogma and large industrial fortunes. Arnold advanced the view that the future of Britain lay in the elevation of its middle classes by means of the culture that had previously been the exclusive possession of its upper

---

[56] Connell 1950: 57–9, 96–7.

[57] See, for example, the speech of Robert Lowe in Hansard's *Parliamentary Debates, 1867,* vol. 188, p. 1549. See also Connell 1950: 61–87.

[58] Arnold, school report for 1852, in Sanford ed. 1889: 19–20.

[59] Arnold 1862a. The reception of Arnold's lectures on Homer is discussed in Coulling 1974: 62–99. On Arnold's "Hellenism," see also Turner 1981: 17–36, 178–80.

[60] Arnold 1865a, 1865b, and 1865c. Arnold's role as a cultural critic is examined in Williams 1958, Eagleton 1984, and Collini 1991.

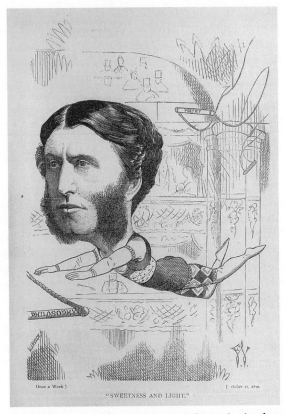

Plate 6. Matthew Arnold, as a trapeze artist, swinging between criticism, poetry, and philosophy (*Once a Week*, 12 October 1872).

classes. He reiterated much of the contemporary rhetoric that eulogized the "middle class" as the bastion of English values – vigor, industriousness, pride, moral integrity. Yet he elaborated at great length upon the cultural shortcomings of that class – its narrowness, tastelessness, intolerance, and ignorance.

Arnold's educational writings were attacked by the Nonconformist press for supporting an interventionist government against precious English liberties and laissez-faire.[61] In contrast, his pronouncements from Oxford had earned him a reputation as a patrician snob and an apologist for established institutions. In an 1867 article in the *Fortnightly Review*, the positivist writer Frederic Harrison satirized Arnoldian

---

[61] The Nonconformist platform that opposed state intervention in education is extensively discussed in Connell 1950: 39–45.

"culture," centered on poetry, and described with terms like "sweet reasonableness" and "light":

> Here are we, in this generation, face to face with the passions of fierce men; parties, sects, races glare in each other's eyes before they spring; death, sin, cruelty stalk amongst us, filling their maws with innocence and youth; humanity passes onwards shuddering through the raging crowd of foul and hungry monsters...and over all sits Culture high aloft with a pouncet-box to spare her senses aught unpleasant, holding no form of creed, but contemplating all with infinite serenity, sweetly chanting snatches from graceful sages and ecstatic monks, crying out the most pretty shame upon the vulgarity, the provinciality, the impropriety of it all.[62]

Harrison, himself a literary critic and journalist, shared Arnold's reservations about the values of England's middle class – its sobriety, machinery, and wealth, its industry, engineering, busy hands, and numbers. He too claimed that literary men could edify and refine this class through their writing and could heal the social divisions in English society through the construction of a universal high culture. As a leading proponent of Comtien positivism in Britain, however, Harrison also held that a *purely* literary culture was useless as applied to action. To counter the baseness and strife governing most Victorian lives, culture needed "principles" derived from a "definite logical process...from history...from consciousness, or from experiment" – in short, a positive science.[63]

Arnold's portrait of middle-class philistinism, as drawn in his last Oxford address, and then elaborated in the four-part essay *Culture and Anarchy*, served to consolidate his quite disparate critics under the rubric of "the enemies of culture."[64] In the process, Arnold also struck an alliance with a man of science. Arnold remarked on a recent editorial in a Nonconformist newspaper that had criticized Huxley for purporting to cure social ills without religion. Although best known for his sparring matches with high churchmen like Samuel Wilberforce, Huxley had also faced charges of materialism and irreligion from evangelical and Dissenting circles since he first began to lecture on the relations of humans and apes.[65] Arnold came to Huxley's defense, asking how a religion "so unlovely, so unattractive, so incomplete, so narrow, so far

---

[62] F. Harrison 1867: 604–5, 610.

[63] F. Harrison 1867: 605, 608.

[64] Arnold 1867 and 1868a. Arnold's response to Harrison and other critics is discussed in Coulling 1974: 181–216.

[65] See, for example, the criticisms from the Scottish press of Huxley's 1862 lectures "On the Relations of Man to the Lower Animals" delivered at the Philosophical Institute of Edinburgh, reprinted in Burkhardt and Smith eds. 1985–2001, 10: 685–97.

removed from a true and satisfying ideal of human perfection" as Protestant Dissent was "to conquer and transform all this vice and hideousness?"[66] Arnold's comments marked the beginning of a long friendship and working alliance with Huxley against the proponents of denominationalism and laissez-faire in education. The clarity, detachment, and critical facility of science, emphasized in Huxley's platform of scientific education as mental discipline, were recruited against the encroaching demand to be practical, progressive, and political. Moreover, association with a paragon of high culture and defender of venerable institutions such as Arnold could deflect some of the criticisms and suspicions that dogged Huxley's own reform efforts.

In a letter to his mother in July 1867, Arnold referred to a discussion he had had with Huxley at the Athenaeum about the education of Arnold's eldest son, who was "getting into debt and not doing well" at Rugby.[67] Arnold considered enrolling the boy in the new International College at Spring Grove. Huxley was a governor of the college and had helped to draft its science curriculum.[68] The college had only recently opened; its sponsors included William Cobden, the Manchester Member of Parliament and former leader of the Anti-Corn Law League, and William Ellis, a wealthy insurance underwriter. An expensive boarding school, largely on the classical model but offering modern sciences and languages alongside more traditional Greek and Latin subjects, the International College seemed to answer Arnold's plea for the enculturation of middle-class philistines. By incorporating subjects that were still widely associated with a practical education into the traditional curriculum of elites, such an institution could elevate the status of those practical subjects, as well as the standing of the pupils who studied them.

Shortly after meeting and befriending Huxley, Arnold began to incorporate elements of science into his own program of cultural reform. In *Culture and Anarchy*, first published in *Cornhill Magazine* between January and July 1868, Arnold presented culture as in part motivated by the "scientific passion," or "the desire after things of the mind simply for their own sakes and for the pleasure of seeing them as they are." Culture, he added, was also underpinned by "the moral and social passion for doing good."[69] Arnold gave these two forces a historical cast as, respectively, Hellenism and Hebraism, and urged that the future progress of England lay in combining them. In the previous year, Arnold had been

---

[66] Arnold 1867: 46.
[67] Huxley, diary entry for April 26, 1867, HP: 70.8. Arnold to Mary Penrose Arnold, 16 July 1867, in Lang ed. 1996–2001, 3: 160. See also Roos 1979: 196–8.
[68] On the International College at Spring Grove, see Bibby 1959: 168–70.
[69] Arnold 1868a: 91.

sent abroad by the Taunton Commission, a second government inquiry on schools, to survey education systems on the continent. The result-ing report, published in 1868 as *Schools and Universities on the Continent*, praised the German system, in which science had a prominent place, as a model for the reform of English institutions. English schools and univer-sities, Arnold remarked, not only failed to teach science but also fostered an "indisposition and incapacity for science." Arnold presented himself as a mediator between "humanists," or proponents of "old classical studies," and "realists," or proponents of "useful studies." To reconcile these conflicting views, he proposed a division of the aims of education into self-knowledge and knowledge of the world, and a corresponding division of subject matter into the "capabilities and performances of the human spirit" (the study of letters), and the "operation of non-human forces, of human limitation" (the study of nature).[70]

In proposing a scheme for combining classical and useful studies, Arnold was concerned not only to integrate science and literature but to provide a means of reducing social tensions and prejudices that were themselves rooted in education. The Taunton Report, to which Arnold contributed in 1867, indicated that education served as an increasingly crucial marker of social status. The report classified English schools based on leaving age and curriculum. "First grade" schools, with a leav-ing age of eighteen or nineteen, were, for the most part, the costly "public schools" of English fame, although there were a few less expensive day schools in the same class. Parents of pupils who attended these schools wished to retain the classical subjects, "for the value at present assigned to them in English society," and would "not wish to have what might be more readily converted into money, if in any degree it tended to let their children sink in the social scale."[71] Parents who sent their children to "second grade" schools, with a leaving age of sixteen, were much less in favor of the classical curriculum. They desired "a certain amount of thor-ough knowledge of those subjects which can be turned to practical use in business – English . . . the rudiments of mathematics, in some cases nat-ural science, or in some cases a modern language . . . they will not allow any culture, however valuable otherwise, to take the place of these."[72]

With its grading system and classification scheme, the Taunton Report tried to encompass the diversity of English schools within a single frame-work, roughly corresponding to conventional social divisions, such as the gentlemanly, and the commercial or professional. Yet the opinions recorded in the report suggest that there were in fact competing models

---

[70] Arnold 1868b: 264, 289–97, 309–18. See also Sanford ed. 1889: 200, 236–7.
[71] *Parliamentary Papers, 1867–8*, vol. 28, pp. 15–18.
[72] *Parliamentary Papers, 1867–8*, vol. 28, pp. 19–20.

of education, each reflecting a different social scale. In his last Oxford lecture, Arnold expressed his concern that culture not be "an engine of social and class distinction, separating its holder, like a badge or title, from other people who have not got it." Defending his model of education against charges of exclusiveness, he went so far as to describe men of culture as "the true apostles of equality."[73] Similarly Huxley, in his address of the following year, "A Liberal Education and Where to Find It," distanced himself from the various parties and interests who currently vied over educational reform, criticizing among others, the richest public schools for training students in "strong class feeling." Like Arnold, he urged state intervention in education along continental lines: "In Germany the universities are exactly what...the English universities are not...corporations 'of learned men devoting their lives to the cultivation of science.'"[74] He added that literature too deserved an important place in a liberal education, properly cultivated into a "refined taste by attention to sound criticism."

In February 1868, Huxley invited Arnold to give an after-dinner speech at the anniversary meeting of the Geological Society of London, at which Huxley was to be installed as president. The seating plan placed Arnold next to Robert Lowe, who, as vice president of the Committee of Council on Education from 1859 to 1865, had been one of Arnold's chief antagonists.[75] Arnold's goals for an expanded school inspectorate had been threatened by economizing measures introduced by Lowe in 1861. Lowe's scheme, popularly known as "payment by results," was designed to reward schools for the performance of students on standardized examinations. According to Arnold, such purely "mechanical" measures would simply increase the prevalence of rote learning and cramming and reduce the role of inspectors to that of clerks.[76] During his tenure on the Committee of Council, Lowe had also argued against government expenditure on classical education, promoting instead the extension of practically oriented science instruction. In an 1864 address to the Philosophical Institute of Edinburgh he had remarked, "I think it will be admitted by all who hear me that as we live in a universe of things, and not of words...it is more important to know where the liver is situated, and what are the principles which affect its healthy action, than to know that it is called jecur in Latin."[77] Lowe's rhetoric of

[73] Arnold 1867: 37, 53.

[74] T. H. Huxley 1868a: 107.

[75] In the event, Arnold declined to speak and the original seating arrangement was changed. See Arnold's letter to Mary Penrose Arnold, 22 February 1868, in C. Lang ed. 1996–2001, 3: 233.

[76] Arnold 1862b, 1862c, and 1862d. Lowe's efforts to establish the scheme of "payment by results" and Arnold's opposition, are discussed in Connell 1950: 203–42.

[77] Robert Lowe 1867: 113.

words and things vitiated Huxley's own polemical discourse of words and facts. It divided culture into the intellectual and practical and positioned the sciences among the latter. On an institutional level, Lowe had sought to use the power of the state to implement a standardized system of education at a minium of expenditure. He tended to weigh the merits of culture in economic terms, and viewed the investment in extending liberal education to be too costly. Those who could afford a liberal education on the classical model could obtain it at a privately endowed school and university. Such a dual system of culture would perpetuate social divisions and relegate the sciences to the status of useful knowledge.

That Huxley should have chosen such an occasion to make a public display of his friendship for Arnold and offer him, as guest of the president-elect of a distinguished scientific society, a platform on which to air his views on education before one of his chief combatants, suggests the importance of their alliance in promoting the expanded role of the state in education and in opposing arguments for free enterprise in the domain of culture. In opposition to Lowe, to other proponents of technical education, to many Nonconformists, and to some defenders of the classics, Huxley and Arnold sought to introduce a single model of culture in both public and private insitutions. They sought thereby to detach culture from any associations with social class, religious denomination, or political or economical interest. They presented themselves in turn as men who were elevated by culture above such divisions and interests and as educators whose role was to bridge these divisions with culture. In their own schemes for educational reform, science and literature were to serve as the foundations of British society, made available at all levels to all persons, but in different degrees. Both men supported measures designed to enable individuals to rise within the system, such as optional Latin and the extension of scholarships to include students of the sciences and modern literature. Their object was not, however, to eliminate inequalities of wealth, or differences of rank, but to create a domain in which such differences were unimportant. Nor was their aim to make all pupils into men of science or letters. But through exposure to certain classic texts of English literature, a few Latin or French authors, a poem learned by heart, and basic principles of physical and natural science, students would acquire an appreciation of finer things and above all a recognition that the possession of a liberal education was a mark of social and mental distinction.[78]

Throughout the late 1860s and the 1870s, Arnold and Huxley met regularly at the Athenaeum, corresponded, exchanged their latest books, and commented on each other's work. In May 1870, Arnold sent Huxley

[78] See, for example, Arnold's school report for 1871, in Sanford ed. 1889: 156–61, and Arnold 1868b: 289–302.

a copy of *St. Paul and Protestantism*, and commented, "I have been reading your Descartes lecture with so much sympathy that I am impelled to do what I can to make you see what I meant and wished in this book I send you."[79] The address on Descartes to which Arnold referred was in part a sweeping plea for "physical science" and a "purely mechanical view of vital phenomena."[80] Yet Huxley, particularly concerned to defend himself against recent charges of materialism, was careful to underscore that mechanism, as a philosophy, and materialism, as a method, were both products of mind. Thought, Huxley claimed, made the human machine "capable of adjusting itself within certain limits," and none were better equipped to perform such adjustments than those who knew the mental mechanism (or was it organism?) best. "The thoughts of men," Huxley suggested, "seem . . . to be comparable to the leaves, flowers, and fruit upon the innumerable branches of a few great stems . . . These stems bear the names of the half-a-dozen men, endowed with intellects of heroic force and clearness."[81] Using an organic metaphor often deployed to describe classical works of literature, Huxley counseled the "Christian young" not to lose faith, but to temper the piety that rendered a "calm pursuit of truth difficult or impossible" and to show "gratitude and reverence" for the "living men, who, a couple of centuries hence, will be remembered as Descartes is now, because they have produced great thoughts which will live and grow as long as mankind lasts."[82]

Huxley's Descartes address, delivered to the Cambridge YMCA, presented science as a branch of "extra-Christian" culture, whose growth entailed the reform of Christianity, relieving it of outworn theology and bibliolatry. That Arnold should have read it "with so much sympathy," is not surprising, for this was the very campaign that he was engaged in with respect to Nonconformists. Arnold had written his mother in 1864, "I mean . . . to deliver the middle class out of the hands of their Dissenting ministers."[83] Though he tended to favor High Church rituals as lending beauty and grandeur to religion, Arnold joined with Huxley in criticizing dogmatic Christianity.[84] The book that Arnold sent Huxley in May 1870, *St. Paul and Protestantism*, purported to verify religious teachings with "reason" and the "facts of experience." Arnold praised Paul's message because of its "immense scientific superiority" and because, unlike

---

[79] Arnold to Huxley, 10 May 1870, in C. Lang ed. 1996–2001, 3: 412.

[80] Huxley 1870a: 181.

[81] Huxley 1870a: 166–7.

[82] Huxley 1870a: 197–8.

[83] Arnold to Mary Penrose Arnold, 16 February 1864, in C. Lang ed. 1996–2001, 2: 282.

[84] Livingston 1988: 30–5.

"Puritan" religion, it was neither "rigid" nor "complete," but "a product of nature, which has grown to be what it is and which will grow more."[85] As Arnold remarked in his letter to Huxley, the book, which was addressed chiefly to the patrons of the schools that Arnold inspected and to opponents of state intervention in education, was an effort to show that the Bible was a part of culture, and not the enemy of culture.

> Conversance... with these singular but powerful people the Protestant Dissenters and... experience [of] how they imagined themselves in possession of an instrument called the gospel which enabled them to regard *de haut en bas* poetry, philosophy, science and spiritual effort of all kinds other than the gospel, to judge it and to do without it, drove me at last to try and show them that the gospel itself was not what they imagined... but that it... was a thing growing naturally and with many parts which must fall away from it or be transformed, and not to be comprehended rightly so long as it is isolated as they isolate it.[86]

Huxley remarked about Arnold's book, "What you say near the end about science gradually conquering the materialism of popular religion... is profoundly true... These people [the Dissenters] are for the most part mere idolaters with a Bible-fetish, who urgently stand in need of conversion by Extra-christian Missionaries."[87] Subsequent correspondence hinted at how far Christianity would have to grow to complete what Arnold and Huxley sometimes called a "new Reformation." Jesus of Nazareth would lie down with Baruch Spinoza. The Bible would become a work of literature and moral science. Christian symbols would take their proper place in intellectual culture, removed from the custody of the closed and ignorant.[88]

In their letters and educational writings, Huxley and Arnold passed swiftly from materialism to material religion to philistine middle class. Through a series of deft conflations, they consolidated their foe, elevated themselves above class affiliations, and made a plea for educational reform. Their easy slippage from "literal," "sensible," and "external" to "practical," "commercial," and "industrial," from "mechanical," "inflexible," and "dogmatic" to "partisan," "sectarian," and "middle class," served to define men of intellectual culture and their open, disinterested, unworldly pursuit of ideal, universal truth. Religion without

---

[85] Arnold 1870: 29–30, 111.

[86] Arnold to Huxley, 10 May 1870, in C. Lang, ed. 1996–2001, 3: 412.

[87] Huxley to Arnold, 10 May 1870, in L. Huxley ed. 1900, 1: 329.

[88] On Jesus and Spinoza, see Arnold to Huxley, 8 December 1875, in C. Lang, ed. 1996–2001, 4: 290–1. On the "new Reformation" see Arnold to Huxley, 13 February 1873, in C. Lang, ed. 1996–2001, 4: 143–4, Huxley's letter to his wife, 8 August 1873, in L. Huxley ed. 1900, 1: 397, and T. H. Huxley 1890f: 17.

high culture could not elevate the English because it suffered from "under-culture," "sectarianism," "intolerance," and a "machinery of covenants, conditions, bargains and parties-contractors such as could have proceeded from no one but the born Anglo-Saxon man of business."[89] In defining a broad and universal culture, which raised individuals above narrow interests and occupations, Arnold and Huxley helped to shape the meaning of middle-classness. The model of science and literature that they promoted contained an image of the new commercial and industrial professions, vast reading audiences, and consuming publics as uncultured, self-interested members of a "middle class" that was very much in need of the kind of broad-minded culture produced by elites.

Neither their friendship nor their common agenda prevented Huxley and Arnold from expressing considerable disagreement on the subject of culture. During the same period in which they were working allies in the field of educational reform, Huxley and Arnold had public exchanges that took the form of disputes. In such disputes, Huxley and Arnold sometimes referred to each other personally as exemplifying opposite positions on the comparative value of science and literature in culture. In his addresses on education, Huxley juxtaposed science and literature in a manner that subordinated the latter and thereby threatened Arnold's exalted view of literary culture. Literature, Huxley asserted, was "neither moral nor intellectual"; the "aesthetic faculty" needed to be "roused, directed and cultivated" by science; and literary culture, while imparting a "sense of beauty" and "power of expression," was unable to furnish a "criterion of beauty" or "anything to say beyond a hash of other peoples' opinions."[90] In his 1880 address at the opening of Josiah Mason's College at Birmingham, Huxley held forth against "classical scholars" who, in ecclesiastical manner, "excommunicated" science "in their capacity of Levites in charge of the ark of culture." Acknowledging Arnold's "generous catholicity of spirit" and "true sympathy with scientific thought," Huxley yet maintained that some of his friend's writings had abetted the cause of these enemies of science: "one may cull from one and another of those epistles to the Philistines . . . sentences which lend them some support."[91]

Arnold himself had engaged in similar tactics for the past fifteen years. From the mid-1860s, he had produced his own self-justifying dichotomies of science and literature, replacing Huxley's hierarchy of "word and fact" with one of "fact and value." While men of letters could

---

[89] See Arnold 1863–4: 318, and Arnold 1870: 14.
[90] T. H. Huxley 1874: 205–6 and 1869: 127–31.
[91] T. H. Huxley 1880: 137, 142, 151–2.

acquire scientific habits, their special access to the moral wisdom in books and to the springs of action in words made their culture superior. Only literature, Arnold wrote, conveyed "high and noble principles of action" and inspired "the emotion so helpful in making principles operative." While the "study of letters" revealed "human freedom and activity," the "study of nature" evinced only "human limitation and passivity."[92] Wanting in moral criteria and active principles, science lacked even the power of "observation" to analyze poetry. Its own "fruitful use," he concluded, depended "on having effected in the whole man, by means of letters, a rise in what the political economists call the standard of life."[93] In 1882, Arnold delivered an address, "Literature and Science," that cast Huxley in the role of an antagonist who trivialized letters.[94]

In the letters between Huxley and Arnold, there is hardly a hint at their extended public exchange. Arnold wrote to Huxley following the Mason's College address, thanking Huxley for the "abundantly kind" remarks; he expressed no desire to "enter into controversy."[95] Evidently, their public dispute was not a matter that came between them in private or that threatened their shared agenda. Scholars have puzzled over their friendship and dispute, seeking to determine which was more fundamental. Some have regarded the common ground between Huxley and Arnold as superficial, or as primarily "personal" rather than "intellectual."[96] But it is clear that fundamental differences and tensions were woven into the fabric of their social identities as men of science and letters. Such differences were in fact productive in the domain of educational reform. The common educational agenda and shared identity as cultural elites that bound the two men together, and that served as a basis from which differences and disagreements could be aired, has been obscured by persistent portrayals of Arnold as the Oxford coxcomb and Huxley as the plebeian.[97] Yet it was in the regal rooms of the Athenaeum, where men of science and letters could mingle over brandy and pipes, that the two first met, and regularly dined. This was not a place for journalists, engineers, workers, or women.

---

[92] Arnold 1868b: 289–92.

[93] Arnold, school reports for 1874 and 1876, in Sanford ed. 1889: 178, 200. Arnold was more explicit about the Philistine character of science in "A Speech at Eton." See Arnold 1879: 21–2, 35.

[94] Arnold 1882.

[95] Arnold to Huxley, 17 October 1880, in C. Lang, ed. 1996–2001, 5: 116–17.

[96] The essentially personal nature of their affinity is argued by Roos 1979. On the superficiality of their agreement, see Irvine 1959: 283, and Coulling 1974: 281–6. Their "intellectual" differences have been stressed by Paradis 1978: 165–6, 194–5, and Super 1977: 233–40.

[97] See, for example, Desmond 1998: 378, 512–13, 631, 639.

It was a place where boundaries were drawn between the makers of a universal culture and the divided masses and classes who needed to be educated. Similarly, at the Royal Academy of Arts, where men of science and letters joined with heads of state, a poet like Arnold was honored as an apostle of reason and light, and a zoologist like Huxley was celebrated as ranking among the foremost writers and orators of the age.[98] By dividing culture exclusively between science and literature, Huxley and Arnold authorized their joint possession of its terrain. In their public confrontations, they shaped the man of science as the converse of the man of letters, apportioning between themselves the values and virtues that were essential to the education of the people. Together, they joined in toasting the royal family, the army, and the navy, the "little realm controlling a vast empire," their "kingdoms of Intelligence," their "conquests" ... and themselves, as the doyens of "Science and Literature."[99]

Scientific Imagination

While still a boy, Huxley had copied passages into his journal from Carlyle's "Characteristics" on "moral greatness," the secret of "genius," and the "region of meditation" below "conscious discourse."[100] Carlyle had used a series of dichotomies to describe the identity of the romantic artist, juxtaposing the "Man of Logic" and "Reasoner" with the "Man of Insight" and "Discoverer."[101] Scholars such as James Paradis, who have noted the importance of romantic and elitist models of science for Huxley, have tended to confine such "influences" to his youth, appropriate to a period of intellectual isolation, social alienation, and brooding. Paradis argues that such models were largely abandoned in maturity, once Huxley acquired an institutional post and threw himself into teaching, administration, and reform. Thereafter, the image that Huxley presented of the scientific practitioner was more utilitarian and democratic.[102] Highly quotable remarks such as "science is nothing but trained and organised common sense" have lent support to accounts of Huxley as an intrepid populist who brought science out of exclusive

---

[98] Huxley contributed to symposia "How to Become an Orator" (Huxley 1888c) and "Good Writing" in the *Pall Mall Gazette*. Printed copies are in HP: 49.55–60.

[99] "Royal Academy Anniversary Banquet," as reported in the *Times*, 2 May 1881, and 5 May 1883. See Roos 1977.

[100] "Thoughts and Doings," entry for April 1842, HP: 31.169.

[101] Carlyle 1831: 352–4, 363.

[102] Paradis 1978: 17–19, 25–7.

places of learning and made it accessible to a wider public.[103] In Desmond's biography, Huxley's enduring reservations about an exclusively utilitarian order of nature and society are interpreted as the product of a residual romanticism, which persisted in the psyche after its moment in history had passed.[104] The mature Huxley's commonsense view of science carries the day, and proves a highly effective ideology for England's rising industrial middle class. Romantic and elitist models of the learned life, borrowed and reworked from writers like Carlyle, were not residual, however, but central throughout Huxley's career. Through accounts of scientific imagination such models persisted alongside other, more populist and utilitarian, forms and served both to identify men of science with the more well-established literati as joint producers of high culture and to carve out a specific role for science within this culture.

In claiming a place for science in the periodical press, Huxley insisted on the value of experiment and observation as grounds for authority. In advancing the cause of science in education, he presented the laboratory and object lesson as superior forms of teaching. In each instance, he built his case in opposition to literature, conceived as a culture of words, whether in the form of journalism, books, classical languages, or college tutorials. He employed an epistemology of austere realism, in which words were at best a means of conveying information and at worst a veil between the knowing subject and the world of facts. This insistence on the transparency of language fit well with his campaign to reform scientific conduct, sweeping away gentlemanly manners by plain speaking. But just as men of letters drew on scientific materials and methods to reshape their own identity as scholars and critics during the period, so too did men of science appropriate aspects of literary identity, such as imagination, in their efforts to inscribe science within culture.

An emphasis on the role of creative imagination in the production of scientific knowledge was increasingly common in the mid-Victorian period. Underlying this were psychological and aesthetic theories incorporating romantic conceptions of genius and invention into the sciences. In works by Herbert Spencer and William Carpenter, for example, ideas were no longer described as "copies" of sensations, as they had been for earlier British philosophers of mind in the empiricist tradition; nor were sensations "copies" of the external world.[105] To the more automatic functions of mind, which received and recorded impressions,

---

[103] T. H. Huxley 1854c: 45–6. For accounts of Huxley as a populist, see, for example, Bibby 1959, Block 1986, and Jensen 1991.

[104] Desmond 1998: 200, 244, 624–5.

[105] See, for example, J. Mill 1829, 1: 40–1.

was superadded the "constructive" or "creative" power of imagination – a synthetic faculty able to unify experience, to constitute objects from concrete individuals, and to abstract from these objects relations not immediately given.[106] Scientific discovery relied heavily on these synthetic powers of mind. As John Tyndall stated before the British Association in 1870, "we are gifted with the power of Imagination . . . and by this power we can lighten the darkness which surrounds the world of the senses . . . Bounded and conditioned by cooperant Reason, imagination becomes the mightiest instrument of the physical discoverer."[107]

Men of science often employed such accounts of imagination in attempts to establish who among the purveyors of culture was more original, and who, therefore, deserved first rank in a high culture pecking order. After rendering imagination a divine "Spirit brooding over chaos," Tyndall assured his audience that in "scientific use" it was "brought before the bar of disciplined reason, and there justified or condemned."[108] Spencer also discriminated between modes of creation, and explained why the symbols of science were not only different from those of literature, but better and truer. In the second edition of his *Principles of Psychology*, he described how the advance of invention was accompanied by a growing sense of the contrast between fact and fiction. The danger posed by arbitrary symbols abated naturally as the creative faculty evolved:

> This progress in representativeness of thought, which brings with it conceptions more general and more abstract, which opens the way to conceptions of uniformity and law, which simultaneously raises up ideas of exact and ascertained fact, which so makes possible the practice of deliberate examination and verification, and which at the same time helps to change belief that is sudden and fixed into belief less quickly formed and more modifiable; is a development of what we commonly call imagination.[109]

Thus literary imagination with its fancy for the concrete came to represent merely the surface of things. In Spencer's psychology, language itself evolved toward abstraction, and what was most remote from sensible experience – the language of science – came to seem most definite and true.

Huxley supported similar views on science and imagination extensively in later works. In a speech delivered near the end of his career, he attacked the precept requiring "the scientific enquirer . . . to abstain

---

[106] On the creative and constructive powers of imagination, see Spencer 1870–2, 2: 529–38, and Carpenter 1874: 487–513.
[107] Tyndall 1870: 130–1.
[108] Tyndall 1870: 133, 165.
[109] Spencer 1870–2, 2: 537.

from going beyond ... observed facts" as a "popular delusion." Science, Huxley asserted, was not a mechanical extraction of general principles from the world of sense. It progressed by means of "the invention of hypotheses," which had "very little foundation to start with." Its theories were not literal truths, but "symbols," part of "that algebra by which we interpret Nature, as if it were absolutely true." Creation in science was impelled by the "divine afflatus of the truth-seeker," a muse that could animate only the wise few. "In science, as in art ... there may be wisdom in a multitude of counsellors, but it is only in one or two of them ... it is to that one or two that we must look for light and guidance."[110] The defining characteristics of Huxley's heroes – Darwin, Newton, Faraday – were the traits of poets and musicians. "The greatest men of science," he wrote, "have always been artists. On the other hand there is ... hardly a great artist who is not in the broad sense of the word a man of science."[111] By claiming originality and genius for the disciplined investigator and experimenter, Huxley effaced the Carlylean dichotomy, which implicitly classed the scientific practitioner together with other narrow, routine producers, efficiency experts, and calculators of a bleak industrial future.

Conclusion: One Culture or Two?

The Victorian controversy over the respective status of science and literature has sometimes been viewed as the beginning of the "two cultures" dispute. C. P. Snow's famous essays, for example, sought to critique the disciplinary boundaries of twentieth-century academic institutions and looked back nostalgically to a time when artists, scholars, and scientists spoke a common language. From this perspective, Huxley appears at the forefront of a movement that would culminate in the separation of the sciences and humanities.[112] It could as easily be argued, however, that Huxley was the architect of *one* culture, not two. In his educational campaigns and his debates with Matthew Arnold and other men of letters, Huxley worked to build a universal "high culture" in which communities of learned men could play complementary roles as educators of a vast and highly differentiated populace. This was not a model of "culture" centered on academic disciplines, but a moral model, in which certain Victorian values, embodied in learned works and practices, could be imparted through study and emulation. The sources and

---

[110] Huxley 1887c: 56–8, 62–4.
[111] Huxley 1882b: 178. See also, HP: 47.147–8.
[112] See especially Stefan Collini's introduction to C. P. Snow's classic essays (Snow 1993). See also Super 1977.

production of this "culture" were exclusionary: it was asserted that only
a few would ever be able to *make* culture, although many could acquire
the means of appreciating it, and of benefiting from it. It is ironic, in light
of Snow's account, that literature and the sciences attained the status of
academic disciplines in England only insofar as they were fashioned
into parts of this "high culture." Many of the differences between scientific and literary practice that Snow bemoans were in fact built into
this model of universal culture and were crucial to its success in the educational sphere. Such divisions and tensions helped to lend authority
to science and literature, by enabling their practitioners to claim that
between themselves they could educate the whole person – rationally,
emotionally, and morally.

In the past several decades, accounts of science and literature in the
Victorian period have been informed by assumptions about a single,
overarching culture, embracing and subsuming differences of profession and specialism.[113] These assumptions have drawn support, in turn,
from discourse models of intellectual history, on the one hand, and from
a range of contextualist approaches to science, on the other.[114] To give,
for example, a literary reading of a text widely recognized as "scientific"
is to cross the boundaries of modern academic disciplines and to open
an interpretive path that was perhaps more accessible to the Victorians. But it is also, potentially, to convert the Victorian construction of
"high culture" into a methodological principle for understanding science and literature. Because both scientific and literary practices were
built into the Victorian construct of culture, interpretations that cross between the science and literature of the period may actually never leave
this exclusive domain and, while noting the tensions between different
discourses, fail to acknowledge that men of letters and science also had
similar agendas.[115]

In this chapter, the science and literature of the Victorian period
have been approached not as aspects of an interdisciplinary or predisci-
plinary "culture," nor primarily as a set of discrete though interwoven
"discourses," but as resources for the forging of identities and communities.[116] An analogous approach has been that applied by British

[113] See, for example, Beer 1983, Levine, ed., 1987: 3–32, Shuttleworth 1996, and Shaffer,
ed. 1998. See also the critique of the one culture model in Small 1994.

[114] See, for example, LaCapra and Kaplan eds. 1987, Christie and Shuttleworth eds. 1989,
and Dear ed. 1991.

[115] For a use of literary methods to indicate the limits of English (and European) cultural
frameworks, see Beer 1996.

[116] Such an approach has been widely applied for readers. See, for example, Fish 1980,
Chartier 1987, and J. Secord 2000.

historians to the notion of "class," by focusing on the variety of ways in which contemporaries used the term and by questioning the explanatory value of this category beyond these more specific and local uses.[117] From this perspective, the question of whether the Victorians possessed one culture or two could be reformulated: what purposes did "culture" serve for learned groups? Men of science, specialist practitioners, and critics emerged alongside men of letters, popular writers, and novelists in the middle decades of the century. The periodical was an important vehicle by which scientific and literary communities were forged through the crafting of different kinds of authority, expertise, and style. As Huxley's early experiences as a review writer show, the points of identity and difference between these groups were still very much in dispute, as was the comparative currency of science and literature for Victorian readers. In the 1860s and 1870s, educational reform became a platform for the construction of a universal "culture." As a polemical construct, culture functioned to advance the status of men of science and men of letters, ultimately by creating positions within higher education for laboratory practitioners as well as literary critics. The very exclusivity of culture assured scientific practitioners a social status above that of the mere technician, and men of letters a status above the journalist. It brought them security against commercial demands, and against the dictates of the literary market. But in so doing, it brought them into confrontation with another learned elite, with deep roots in the English schools and universities: the Anglican clergy. As the following chapter will explore, the culture of science, as it was established in education, in the periodical press, and in Victorian homes, was also a religious culture.

[117] See, for example, G. S. Jones 1983, Joyce 1991, and Wahrman 1995.

4

The Worship of Science

A tendency to excessive reverence for men of science . . . often subdues me, and, when I find myself unsustained in my inmost convictions, depresses and afflicts me.[1]

– James Martineau, 1868

If Huxley's scientific identity was derived in part from the identities of the artist and man of letters, how then was the man of science defined in relation to the clergyman, the figure who in many respects appears his most obvious counterpart or rival? Until the early Victorian period, scientific practitioners had often been clergymen by vocation. Leading naturalists such as William Buckland, Adam Sedgwick, and John Henslow not only combined their respective geological and botanical pursuits with clerical office but actively incorporated their science within the Anglican tradition.[2] Natural theology continued to provide a unifying structure for English science well into the nineteenth century.[3] Even practitioners like Darwin, who chose not to pursue a Church living, and whose work was viewed by some as undermining the principle of design in nature, still occupied traditional positions in society that

---

[1] Letter to Frances Power Cobbe, 18 November 1868, Frances Power Cobbe Papers, Huntington Library, CB 592.

[2] On Buckland, Sedgwick, and the Anglican tradition of geology, see Gillispie 1951 and Rupke 1983. On Henslow, see Russell-Gebbett 1977 and Browne 1995.

[3] The importance of natural theology as a framework for English science and the large literature on the topic are discussed in Brooke 1991.

rested partly on religious foundations and exercised a quasi-religious authority in their local communities.[4] The appearance of Darwin's *Origin of Species* and other works, however, has long been associated with the decline of religious authority over the mind, and as evidence of the gradual secularization of knowledge and the corresponding replacement of religious leadership in a variety of quarters by new professions.[5] The view that scientific and religious ideas or beliefs were in a state of unremitting struggle during the mid-Victorian period has now undergone substantial revision.[6] The "conflict" of science and religion has been reinterpreted as a contest between professional groups for cultural authority, or as part of a more general clash between the Anglican privileged elite and the new industrial, largely Nonconformist classes.[7] Such arguments have shifted the emphasis away from the problem of belief and, in place of ideas, have presented socioeconomic criteria as the predominant means of explaining controversies between men of science and clergymen. Such accounts, however, have still adhered to the general model of a fundamental opposition between science and religion during the period.

A second set of revisions of this picture has reintegrated the problem of belief within the history of scientific culture. Debates over evolutionary theory, once taken as the major battleground of the warring camps of religion and science, have been shown to have occurred largely within established theological traditions.[8] Some historians have demonstrated the extent to which new scientific theories could be accommodated within orthodox Anglicanism. Many practitioners, such as the zoologists George Romanes and St. George Mivart, retained strong religious convictions despite their advocacy of human descent from other species.[9] Other work has shown, however, that these new accounts of the natural world were also made to serve radically different supernatural doctrines. Thus, for example, Alfred Russel Wallace argued for a scientific approach to spiritualism, while the notorious proponent of scientific naturalism, John Tyndall, explored the possibility that immaterial or spiritual forces were active in the universe.[10]

---

[4] Moore 1985.

[5] See, for example, Chadwick 1975: 165–75 and Heyck 1982.

[6] The "conflict" interpretation is carefully discussed in Moore 1979; see also Desmond 1998: 632–6.

[7] See MacLeod 1970b, Young 1985, and Turner 1993.

[8] Moore 1979 and 1990.

[9] Turner 1974.

[10] Wallace's spiritualism is examined in Oppenheim 1985. On Tyndall's "pantheism," see Barton 1987.

It is accordingly difficult to view the Victorian period as one in which ecclesiastical power was simply usurped by that of science. The structures of authority available within the Anglican hierarchy provided useful models for men of science, who presented themselves as, and were perceived by many contemporaries to be, a new kind of priesthood, interpreting laws of nature that all must obey or disregard at their peril. The authority of nature thus came to stand alongside that of Scripture and Christian tradition, but in many respects the shape of that authority was modeled on religious forms. Men of science and religion sat together on the boards of public institutions and royal commissions and consequently faced similar problems. In the reform of institutions like the Church or the education system, the need to find collaborative strategies where differences in belief could be set aside structured the ways in which religious forms would be accommodated in scientific conduct. Such pragmatic solutions were also reached within households or friendships, even if highly publicized quarrels involving men of science also reveal that such strategies were contingent and ephemeral.

Huxley, who often employed militaristic rhetoric against established religious authorities in his campaigns for scientific education, has been perhaps the most difficult to accommodate in any single revisionist model. Contemporaries remarked on Huxley's clerical manner: he was "Pope Huxley," the preacher of "lay sermons."[11] Yet Huxley's use of religious formulae and personae has often been taken as a mockery rather than as an outright appropriation. Huxley did indeed engage in frequent and strident debates with Church leaders and declare science the enemy of parsondom.[12] His famous confrontation with Samuel Wilberforce, the bishop of Oxford, over the merits of Darwin's theory of descent, continues to epitomize a clashing of world views and social interests, even while it is acknowledged that there were substantial discrepancies among contemporary reports.[13] However, Huxley too borrowed substantially from religious traditions. His view of Nature as an austere and merciless judge has been described as "scientific Calvinism."[14] His agnostic creed has been likened to various positions articulated by Anglican theologians on the limits of religious knowledge, and his iconoclastic style and reforming enthusiasm have been traced to Dissenting traditions.[15] Huxley also figures prominently in accounts that interpret

---

[11] See Lightman 1983. For a contemporary account, see Hutton 1870.
[12] See, for example, Huxley 1890f: 16–17.
[13] Lucas 1979, Gilley and Loades 1981, Morrell and Thackray 1981, Jensen 1988.
[14] Moore 1979.
[15] Lightman 1987, Desmond 1998: 622–6.

the relationship between science and religion as a social conflict pitting old and new professional groups against one another.[16]

Approached as a problem of identity formation, rather than as one of belief or professionalization, the relations between science and religion appear rather differently. Huxley's efforts to define science as culture, and to increase the prominence of scientific subjects in liberal education, involved considerable work from within Anglican traditions and institutions as well as persistent attempts to distance science from purely commercial and industrial pursuits. This chapter will show how old and new professional groups worked together to consolidate authority, employing scientific and religious forms in conjunction with each other. Huxley might have coined the term "agnosticism," but in the privacy of his own home he could present himself as a religious devout. He also conducted an extensive correspondence with a clergyman, Charles Kingsley, over the immortality of the soul; and, as a member of the London School Board, he defended the use of Scripture in the classroom against secularists. Such activities refute the criterion of secularization as fundamental to the emergence of science as a profession.

Holy Man

In Chapter 1, allusions were made to the importance of religion in the process of self-construction that Huxley undertook with his fiancée, Henrietta Heathorn, during the late 1840s and early 1850s. While Heathorn employed Christian virtues to give value and dignity to her work and position in the household, Huxley tried to mobilize moral qualities associated with religious life for his own calling. Concerned about the differences in their beliefs, Huxley wrote to his fiancée shortly after their engagement, explaining to her how religion was fundamental to his manhood:

> I have thought much of our afternoon's conversation, and I am ill at ease as to the impression I may have left on your mind regarding my sentiments. If there be one fact in a man's character rather than another, which may be taken as a key to the whole – it is the tendency of his religious speculations. Not by any means, is the absolute nature of his opinions in themselves a matter of so much consequence, as the temper and tone of mind which he brings to the inquiry ... May his fellow men then form no judgment upon the point? Surely they must ... but let them ... inquire whether he be

---

[16] See especially Barton 1983, 1990, and 1998. Desmond, too, employs the professionalization model, though frequently evoking broader social determinants (see, for example, Desmond 1998: 334–5, 632–6, and Desmond 2001).

truthful and earnest – or vain and talkative – whether he be one of those who would spend years of silent investigation in the faint hope of at length finding the truth – or one who conscious of capability would rather gratify a selfish ambition by adopting and defending the first fashionable view suited to his purpose ... On grounds of this kind only can a judgment be justly formed. On these my own dear only must you form your judgment of me.[17]

By focusing on the process rather than the substance of belief, Huxley directed discussion away from the more contentious subject of theology. Using the resources of liberal theology and romantic criticism, he was in effect redrawing the boundaries of religion. By relocating its essence in ways of thinking, feeling, and conducting oneself with respect to belief, he could express religious devotion through his own special calling, scientific inquiry. Through these boundary negotiations, Huxley tried to defend his unorthodoxy and to challenge the authority of others, particularly Heathorn, to judge him on their terms. What Heathorn expected and required of him was not selfless, protracted, and unrewarded investigations, but conviction and action in the world of work. Through repeated examples of how her own faith gave her strength to overcome obstacles and command others in the home, she confronted him with quite different values and uses of religious life.

In ensuing correspondence, Huxley presented Heathorn with ways in which his concerted practice of inquiry might fit into the world of other men's labors. On returning to London, he depicted the Crystal Palace as "the great Temple of England" where fifty thousand faithless people worshipped every day, while "their professed teachers" disputed over the abstruse theology of the Anglican creed. He remarked how the "profound ragings" of Carlyle's *Latter Day Pamphlets* awakened in him a vision of "that union of deep religious feeling with clear knowledge" by which alone men could live in peace. He hoped that Carlyle's efforts, "and those of such men as he – earnest resolute and singleminded men," would "do something towards producing a thorough change in our public men and public acts."[18] Combining the characteristics of the holy man and the man of action – determination, courage, strength – the man of science could become a prophetic figure, razing outworn traditions and helping to build new moral foundations:

> Of one thing I am more and more convinced – that however painful for oneself this destruction of things that have been holy may be – it is the only hope for a new state of belief ... That a new belief – through which the faith and practice of men shall once more work – is possible and will exist – I

[17] Huxley to Heathorn, 18 October 1847, HH: 3.
[18] Huxley to Heathorn, 17 July 1850, HH: 112.

cannot doubt. Whether it will arise in my time or not is another matter. At any rate, what am I, that I should not be content even by negation to help in the "Forderung der Tag" as the great poet has it?

Think well of these things Menen. I do not say think as I think – but think in my way – Fear no shadows – least of all that great spectre of personal unhappiness which binds half the world to orthodoxy. They say "how shocking, how miserable to be without this or that belief!" Surely this is little better than cowardice and a form of selfishness. The Intellectual perception of truth and the acting up to it – is so far as I know the only meaning of the phrase "one-ness with God." So long as we attain that end does it matter much whether our small selves are happy or miserable?[19]

Here Huxley submitted himself to Heathorn not merely as the silent investigator or discoverer but as the public instructor. As a man of science, he was a moral legislator and educator, shaping the character of others by dictating the very terms according to which they could believe and act. This prophetic persona, felicitous in resolving Huxley's own problems of belief and agency, was invaluable in sealing his relationship with Heathorn. Together with her models of womanhood, his religious identity furnished the couple with a common set of goals and prescribed the very different roles that each could play in fulfilling these aspirations. Thus while Huxley urged Heathorn to "think in my way," he also paid tribute to her "simple and abiding faith" – a form of belief, based upon feelings, that was appropriate to her woman's nature and supportive of his own heroic struggles with the foundations of belief.[20] Heathorn in turn wrote odes to scientific genius and apologies for agnosticism in her married life, endowing these activities with religious significance and accompanying her children to church.[21] A source of great tension between them, religious symbols and practices were also a means of sustaining their affection, sanctifing the very differences that divided them.

The prophetic role that Huxley assumed in private with Heathorn was also important in his building friendships with more orthodox professional men and in how he presented himself to audiences insistent upon religious leadership. In 1859 he wrote to one of his closest friends, Frederick Dyster, a physician and also a friend of the liberal Anglican cleric Charles Kingsley:

As to the methods by which the Biblical writers arrived at their great truths I do believe that they were in the truest and highest sense scientific – I recognize in their truths the results of a long and loving if sorrowful,

[19] Huxley to Heathorn, 23 September 1851, HH: 165–6.
[20] Huxley to Heathorn, 23 September 1851, HH: 165–6.
[21] H. A. Huxley 1913: 11–12, 75–6, 99–101, 155–6.

study of man's nature and relations... Thou shalt love thy neighbour as thyself is the law of Gravitation of society.

Theology and Parsondom... are in my mind the natural and irreconcilable enemies of Science. Few see it but I believe we are on the Eve of a new Reformation and if I have a wish to live 30 yrs, it is to see the God of Science on the necks of her enemies. But the new religion will not be a worship of the intellect alone.[22]

This letter might be read as the kind of declaration of war between science and religion for which Huxley has become famous. Such military rhetoric was appropriate, however, for a prophetic figure summoning the power of his own jealous God in the name of true religion. Rather than simply divide the world into enemy camps, Huxley assimilated Scripture, its authors and prescriptions, by conflating religion and science. This kind of religious borrowing was not mockery, but a way of reconciling differences with a close friend at whose seaside retreat Huxley had once honeymooned, and where he often repaired for holidays of dredging and dissecting. It was also a way of appealing to the liberal wing of a medical profession from whom many men of science, Huxley included, came – the same profession that, in the 1840s, had lionized Richard Owen as a great reformer. Through his use of religious symbols Huxley tried to sanctify his own reformism, which involved an attack on the theological tradition and on the High Anglican community that a previous generation of practitioners, Owen in particular, had courted.

The religious role that Huxley adopted in his letters to Dyster was continuous with the public image he promoted from the mid-1850s onward. In an 1859 address, reported in the London weekly *The Builder*, he declared all "religious reformations" to be the work of the "scientific spirit" and noted that "the great deeds of philosophers have been less the fruit of their intellect than of the direction of that intellect by an eminently religious tone of mind. Truth has yielded herself rather to their patience, their love, their single-heartedness, and their self-denial, than to their logical acumen."[23] A variety of platforms were set up for such addresses in the Victorian period. The audiences ranged from the upper to the lower middle class, though most events were ostensibly places where all classes could meet – places where class differences could be transcended in a universal Christian discourse of knowledge. Such platforms, erected in town halls, Mechanics' Institutes, and Friday

---

[22] Huxley to Dyster, 30 January 1859, HP: 15.106.

[23] T. H. Huxley 1859d. See also Roos 1979: 170–1. A major journal of engineering and architecture in the middle decades of the nineteenth century, *The Builder* was important for Huxley's patrons and superiors in the School of Mines and the Science and Art Department.

evenings at the Royal Institution, continued a tradition from the 1810s and 1820s of presenting natural knowledge in a religious framework. They were not the sort of popular occasions that men of science avoided in the interests of professionalization. They were prestigious forums for the advancement of a career – stages on which men of science worthy of the title were expected to perform.[24]

In 1866, Huxley appeared with William Carpenter, John Ruskin, Frederick Maurice, and Charles Dickens in a public lecture series titled "Sunday Evenings for the People" at St. Martin's Hall. In a style characteristic of a Puritan sermon, Huxley began his address, "On the Adviseableness of Improving Natural Knowledge," with an account of the punishments that had been visited on seventeenth-century England, the plagues and the great London fire, the deliverance that had been sought through prayers and offerings, and the redemption signaled by the birth of the Royal Society. He described the methods of scientific inquiry, first advanced in this society, and the healing powers and improvements that they had brought. In effect, he offered scientific practices as a body of religious exercises involving devotion, ardor, earnestness, and abstension, which alone could work the miracles for which Christians past and present still hoped and prayed. Science was not authoritarian, like the theology of old. Its truth, he asserted, was based not on the testimony of individuals, texts, feelings, or intuitions (all corrupted by bias, privilege, and custom), but on "evidence" gleaned from disciplined observation and rigorous experiment: "The man of science has learned to believe in justification, not by faith, but by verification."[25]

In later works, the more elaborate of which remained unfinished, Huxley built grand stages for his prophetic persona, rewriting the history of science as a history of religion and vice versa. He cast "doubt" – the power that worked by "patient consideration and reconsideration" – as the chief agent of progress and purification and as the great object of persecution.[26] In a lengthy manuscript on the "History of Christianity," he located this agency in the Hebrew prophets, who reformed Jewish religion from a ritualistic, ecclesiastical system into an ethical one, making divine power conditional upon ethical deeds: "They appeal to the authority neither of the priest, nor of tradition, nor to any written law. Their sole guide is the inner light of reason and conscience." He attributed the success of Christianity to heroic reformers like these, who obeyed their own clear head and conscience "against the world."

[24] On public lectures in science, see Hays 1974 and 1983.

[25] T. H. Huxley 1866: 41.

[26] HP: 45.104–8. This manuscript, part of a projected history of civilization, probably dates from the late 1880s or early 1890s, the period of Huxley's most extensive writing on history and religion.

Setting the scene for the new lights of science, the new rituals of laboratory, lecture room, and learned debate, he credited all great change to "solitary, exceptional men" who were willing to break with tradition and assemble an "energetic minority."[27]

In 1874, Francis Galton conducted a survey according to which 80 percent of English men of science professed to be Christian.[28] The religious identity of men of science depended less on their personal beliefs, however, than on what others expected them to believe and on how they presented matters of belief to others. Approached as a form of representation, the religious identity that Huxley professed so extensively in public and private was clearly not of his own making. It was an acute part of the language of representation that bound men of science as a community, because it was crucial in expressing their relationship to others in mid-Victorian society – crucial, that is, in articulating their social role. Like domesticity and feminine virtue, classical learning and letters, the figure of the holy man was part of the mythology of their genius because it was part of the fabric of social order and authority in Victorian England.

In one of the numerous specialized studies of religion that emerged in the second half-century, Herbert Spencer displayed a fascination with the power of religion as both a social bond and an agent of change. Stitched together from the more technical work of others and from the advice of correspondents, and sustained by contributions from friends and subscribers, Spencer's *Principles of Sociology* maintained that religious institutions and beliefs, past and present, were the foundations of society; that the nature of religion was adapted to the needs of an evolving social state; and that science, as the agent of progress, was the purifier of religion, bringing greater consistency to beliefs and enlarging the sense of wonder and awe among believers.[29] The work gave the texture of sociological analysis to a position that Spencer had put more abstractly both in his *First Principles* of synthetic philosophy and in his essays on education: "Devotion to science is a tacit worship."[30]

In 1879 the writer Moncure Conway solicited men of science for a new Association of Liberal Thinkers, organized for the expressed purpose of reforming religious practices through the application of scientific methods. Conway was a leader of the South Place Religious Society and

---

[27] HP: 48.138, 180–1. This 100-page manuscript, entitled "The History of Christianity," probably dates from the late 1880s or early 1890s and was composed as part of a projected series of "Lectures to Working Men on the Bible."

[28] Galton 1874: 127.

[29] Spencer 1876–96, 3: 3–175.

[30] Spencer 1860–2 and 1861: 41. For a general account of the comparative study of religion that emerged in Britain in the 1860s, see Wheeler-Barclay 1987.

Chapel in Finsbury, and the author of didactic works on Carlyle and Thomas Paine and of moral compendiums like *Wise Men of the East* and *The Rules of Civility*. A circular composed by Conway stated the association's objectives: "The scientific study of religion, the collection and diffusion of information concerning religious developments throughout the world, the emancipation of mankind from the spirit of superstition, fellowship among liberal thinkers of all classes, the promotion of the culture, progress and moral welfare of mankind, and of whatever in the form of religion may tend towards that end."[31] Although the organization soon dissolved, Conway successfully petitioned Huxley to act as president, Tyndall as vice President, and the mathematician William Clifford to be among the council members.

Through their subscription to such projects and societies, men of science asserted their commitment to social progress and liberal culture, constructing a historical role for themselves as purifiers of religion. These systems of knowledge and forms of association, like their private friendships and domestic relations, gave men of science the moral leverage to criticize other faiths that were not based on "evidence" and to conduct themselves as righteous tutors of an unlearned, credulous, and irresponsible public. But could men of science, through their public performances and private relations, shape what their audiences understood by "religion"? The placard announcing Huxley's 1866 address at St. Martin's Hall advertised the Sunday evening series as "discourses on science and the wonders of the universe" designed for the "large numbers of those who at present do not attend places of worship," "producing in their minds a reverence and love of the Deity, and raising up an opposing principle to intemperance and immorality."[32] Evidently the series organizers sought to cultivate an audience disaffected with institutional religion and seeking new (preferably theistic) forms of belief, edification, and diversion. But Huxley's oracular address on the religion of science, which caused the series to be canceled, suggests how men of science could lose that audience.[33] His private negotiations, particularly with Heathorn, show likewise how difficult it might be for men of science to control the meaning of religious symbols in social circles where they were still expected to be natural theologians.

Following his presidential address at the British Association meeting in Aberdeen in 1868, Joseph Hooker wrote to Darwin of the difficulties that even a modest man of science might have in managing a lay audience:

---

[31] The circular was enclosed in Conway's letter to Huxley, 7 January 1879, HP: 12.300–03.

[32] HP: 31.189. This advertisement has been reproduced in Desmond 1998.

[33] See the description of the event in Desmond 1998: 344–6.

What a bother the papers kick up about my mild theology! An Aberdeen one calls me an Atheist and all that is bad: to me, who do not intend to answer their abuse, misquotations, garbled extracts, and blunders, it is all really very good fun. There were gentle disapproving allusions at Kew Church today I am told! I am beginning to feel quite a great man!

Tyndall most assuredly did couple our names most prominently, unequivocally, and unmistakably as the two modestest men in Science!!!

The last day was by far the hardest work, what with Committees and Councils and the Mayor's Dejeuner – Huxley made a sad mess of it by twice offending the clergy (The Clergy throughout behaved splendidly like men and gentlemen. The Cathedral service was glorious, the anthem was chosen for me "what though I know each Herb and Flower" and brought tears into my eyes, and Dr. Magee's discourse was the grandest ever heard by Tyndall, Berkeley, Spottiswood, Hirst or myself) totally without cause or warrant, once at the Prehistory Congress, when he likened them to Bulls of Bashan and again at the Red Lion Club, when they got up and left the room! I was not there having providentially been prevented attending. [34]

In a contemporary report of Hooker's address, it was stated that most of the audience cheered when he was thought to be supporting a scriptural account of creation, while only a small portion applauded when he declared himself opposed to such accounts.[35] In his letter to Darwin, however, Hooker displayed his disregard for the opinions of a lay public but his high regard for a clerical elite. He suggested that relations with the clergy might be managed through manners, that fellowship might be made over religious symbols, even those of the High Church, and that Huxley's immodest proposals, which alienated the clergymen – a community much more widely recognized than men of science to be authorities on matters of religion – were tactless and impolitic. In their debates over religious authority, men of science had rather more in common with members of the clergy than they did with the middle- and working-class publics whose reverence they sought. The prophetic model of the man of science was very close to the patriarchal form of heroism being preached at the time by liberal Anglicans such as Thomas Arnold, Frederick Maurice, and Charles Kingsley, the leading architects of Christian manliness.[36] Liberal theologians like Kingsley, Arthur Stanley, Frederick Farrar, and the authors of *Essays and Reviews*, were also makers of a "new reformation" and notorious critics of the churches for failing to work in the world. Many were also friends.

---

[34] Hooker to Darwin, 30 August 1868, DAR 102: 229–32.
[35] *Daily News*, 21 August 1868, p. 3. Cited in Ellegård 1958: 166–7.
[36] On Christian manliness, see Newsome 1961, Vance 1985, and Hilton 1989.

A Broad Church

In 1868, Huxley was invited by the Reverend James Cranbrook, minister of a Unitarian church in Edinburgh, to speak before his congregation. The resulting work, "On the Physical Basis of Life," was considered by Huxley to be one of his most pointed defenses of materialist methods of inquiry. He concluded with the prediction that men of science using such methods would soon be able to create life from inanimate matter in the laboratory.[37] The address was delivered on Sunday, 8 November, after Huxley had received several appeals from the Scottish cleric emphasizing the urgency of enlightening local opinion about science.[38] It was widely reported, and subsequently published in the *Fortnightly Review*. One report, appearing in the *English Independent*, drew the conclusion that physical force was permeated and directed by spiritual power: "a protoplasm which science cannot detect."[39]

That such a piece, ostensibly anticlerical, was solicited by a clergyman and interpreted theologically suggests that the refashioning of religious models by scientific practitioners was only part of a more intricate social process involving the reconstruction of clerical identities and church communities. Since the 1820s, Anglican clergymen had used developments such as the decline in attendance at Sunday services in poorer urban parishes, the admittance of Catholics and Dissenters to Parliament, and the emergence of new secular institutions such as the University of London, to call for reform within the Church. In the course of extensive debates over Church reform, discourses about scientific method and authority were important in consolidating the "broad" or "liberal" community of Anglicans in the middle decades, enabling them to define their own clerical role as God's instruments in the unfolding of revealed truth.[40]

In 1863, a Privy Council ruling overturned the condemnation of several authors of the liberal theological work *Essays and Reviews* passed by the ecclesiastical Court of Arches.[41] In their defense, the authors made heavy use of the language of scientific inquiry to legitimate their role

---

[37] Huxley 1868b.
[38] Cranbrook to Huxley, 2 and 23 October 1868, HP: 12.328–34.
[39] *English Independent*, 18 February 1869, p. 151. Cited in Ellegård 1958: 316.
[40] Accounts of the "broad church" movement and liberal theology in Victorian England may be found in Sanders 1942, Norman 1976, and Reardon 1982. Links between liberal Anglicans and scientific professionals have been explored in Moore 1990.
[41] On the controversy over *Essays and Reviews*, see Altholz 1994 and Shea and Whitla eds. 2000. For a study of the debates over clerical subscription and the formation of belief, see Livingston 1974.

as biblical critics. In a confessional tract that he wrote during his trial, Rowland Williams stated that a clergyman "binds himself to study" by ordination and "is responsible before Heaven for the task of bringing the Church's doctrines into harmony with fresh facts and wider experience."[42] In support of the authors, J. W. Colenso, the bishop of Natal, himself threatened with excommunication for his work *The Pentateuch and Book of Joshua Critically Examined*, depicted "the science of manuscripts and of settling texts" as God's scourge, and himself as a patriarch "drawn in with the stream," his ark launched "upon the flood" of purifying waters.[43] After the Privy Council ruling, broad churchmen employed similar methodological terms to renew their petition to revise clerical subscription to the Anglican creed. In his letter to the lord bishop of London, Arthur Stanley, then dean of Westminster, wrote that the subscription law "habituates the mind to give careless assent, and leads to sophistry in the interpretation of solemn obligations."[44] In redefining their clerical role, liberal Anglicans thus employed the same tactics that Huxley had used in his private correspondence and public statements on matters of belief. By focusing discussion on the formation of belief rather than on doctrine, broad churchmen appropriated the practices and materials of science to reconstruct their own identity. Their judicial campaigns indicate some of the ways in which scientific practices could be appropriated for religious purposes and could help to shape a clerical community.

In periodicals, review columns, learned societies, and churches during the 1860s and 1870s, men of science and clergymen participated (often side by side) in a series of debates to determine which, among many contending groups, were Victorian England's true priests and prophets. In such debates, Huxley's essay "On the Physical Basis of Life" could be read as a theological text, and new works of biblical interpretation like *Essays and Reviews* could appear as works of science. Topics such as evolutionary theory, free will, spiritualism, materialism, and the efficacy of prayer were woven into these controversies. Underlying these specific issues were crucial questions about the language in which they were cast and the forms of agency that they implied. To describe evolution, for example, as the unfolding of Divine will was to depict a universe whose substance and forces were superintended more appropriately by certain groups than by others; it was to prescribe not only which laws but which lawgivers, texts, and forms of redemption were fundamental.

[42] Rowland Williams 1861–2: 13. On appeals to science by other *Essays and Reviews* authors, see Shea and Whitla eds. 2000: 66–73, 346–8.

[43] Colenso 1862–79: 1: 220.

[44] Stanley 1863: 25.

Not surprisingly, one of the most contentious issues of the period was that of belief itself. Discourses on evidence and verification were invoked by all parties in different ways. In a symposium in the periodical *Nineteenth Century* on morality and religious belief, the Unitarian minister James Martineau defended the position that ethical beliefs were based not on external laws, or mere rational deliberations, but on inner feelings informed by reason and revelation. Such feelings, like those of obligation and freedom of will, were epistemologically equal if not superior to physical sensations and constituted real "evidence" for belief in a personal God and immortality.[45] Arguments similar to Martineau's, conferring the status of evidence on the very entities that men of science often dismissed as subjective – religious "feelings" – were utilized extensively by liberal Anglicans of the period in campaigns against the "external authority" of Church doctrines and institutions. They were also invoked by Church conservatives to resist these same reforms. In the preceding issue of *Nineteenth Century*, William Gladstone had claimed that the "authority" of tradition was simply the wisdom of ages of inquiry, a "scientific principle" that enlarged one's vision. The embrace of inner feeling was indolent, fostering an "ignoble servitude" to the dictates of the popular press.[46]

If theologians could trespass onto allegedly scientific ground, and conduct poaching operations on behalf of Church interests, men of science could not defend their property rights by referring to well-drawn boundaries of scientific language, discipline, and method. When Huxley, Tyndall, or Clifford participated in journal symposiums, claiming that their beliefs were based not on "tradition," but on "evidence and demonstration acting upon reason," they could not rely upon well-defined canons of evidence that were theirs alone to wield.[47] By entering such debates, men of science were engaging in a process of intense negotiation with clergymen and other religious leaders about the meaning of practices taken to be fundamental to both science and religion. Such negotiations were unavoidable, for the beliefs of clergyman and their public pronouncements about belief were crucial in determining the meaning of science in the period. Moreover, the extensive borrowing of scientific practices by clerics was actively encouraged by men of science. The claims of *Essays and Reviews* authors and of exegetes like Gladstone to be following scientific methods actually supported narratives like Huxley's "History of Christianity," in which biblical writers appeared

---

[45] Martineau 1877: 341–4.

[46] Gladstone 1877a: 10, 21; and 1877b: 904.

[47] Clifford 1877: 71. A variety of contributions to this Victorian debate on belief have been collected in McCarthy ed. 1986.

as masters of induction. The attention given to moral and metaphysical issues in sermons, periodicals, and other public forums suggests the importance of such issues as terrain to be possessed by men of science and clerics. Yet these debates were not simply warfare by other means. They were also a process whereby scientific practitioners and clergymen reinforced the boundaries of their own practices and communties and enhanced their common authority as men of learning able to command an audience's assent and reverence.

Debates between men of science and liberal clerics were also carried out in private, where they served as a means of self-definition and of building a common ground in which differences could be discussed and diffused. Such exchanges were not unusual during the 1860s and 1870s, as indicated, for example, by the correspondence of James Martineau with Joseph Hooker on theology.[48] Frances Power Cobbe, a Unitarian philanthropist and religious writer, traded letters with Charles Lyell on free will and with John Tyndall on the efficacy of prayer.[49] Huxley wrote to the liberal Anglicans Benjamin Jowett, Frederick Farrar, and Arthur Stanley on matters of scientific and religious education. His most extended and intimate correspondence, however, was with Charles Kingsley. In a letter to Kingsley in 1860, Huxley described how their mutual friend, Frederick Dyster, had tried to introduce them in 1855, Kingsley being the only person "who would do me any good."[50] In the mid-1850s, Kingsley was still involved in Christian Socialism, a pedagogic movement that had been formed by liberal Anglicans during the Chartist protests in 1848 to counsel England's workers against revolution and to shame England's ruling classes for renouncing paternalism. Together with Frederick Maurice, Thomas Hughes and others, Kingsley wrote novels, sermons, and tracts criticizing the worship of Mammon, and he taught at the Working Men's College, which Christian Socialists had opened in London in the early 1850s.[51] During the same period, Huxley was delivering his first lectures to workers at Friday evening courses in South Kensington. In 1854, he had approached Maurice with an offer to coordinate his natural history teaching with the courses in literature and the social sciences at the Working Men's College.[52]

Correspondence between Huxley and Kingsley began in December 1859 when Kingsley expressed appreciation for a review of the *Origin of Species* that Huxley had written for *Macmillan's Magazine*. Kingsley

[48] Drummond and Upton eds. 1902, 1: 439–40.
[49] Cobbe 1894, 2: 89, 120–3.
[50] Huxley to Kingsley, 23 September 1860, in L. Huxley ed. 1900, 1: 217–22.
[51] On Christian socialism, see Backstrom 1974 and Norman 1985. On the Working Men's College, see J. Harrison 1954.
[52] Huxley to Maurice, 19 December 1854, HP: 200–1.

wrote that he hoped it would "keep the curs from barking till this great question has been fairly discussed by men who really know something about the matter," and he praised Huxley's more technical articles in medical journals: "you among all living English Naturalists may be a source of truth on these things."[53] Shortly after receiving this letter, Huxley wrote to Dyster saying that he found Kingsley "a very real, manly, right-minded person" and "an excellent Darwinian," adding that on the whole "it is more my occasion to correct him than his to correct me."[54] In these initial letters, Huxley and Kingsley addressed each other as allies, each supporting the other's character and learning against incursions from the wild and untrustworthy journalistic rabble. They also seemed to share views about evolutionary theory.

In September 1860, a brief note from Huxley on the death of his first son, Noel, brought a long letter of condolence from Kingsley and gave rise to a series of exchanges over the immortality of the soul, the nature of belief, and the meaning of science and religion. Kingsley offered Huxley consolation through his own intimations of eternity. This "instinct," upon which Kingsley based his belief in the persistence of personality after death, might some day be supported by scientific induction:

> [T]here comes over me at times a vision which I would not sell for all the gold on Earth, of infinite and perpetual upward development, going on all around from all ages to all ages, of the inorganic into the organic, of the organic into the animal, the animal into the rational, the rational into the higher sphere of being for which as yet we have no name save words abused by partial conceptions... the moral – the spiritual – who can define it, though it is life of our life and light of our light.

Kingsley described how his religious feelings gave him a sense of command over the natural laws of life and death and how his belief in evolutionary theory was premised on the same certainty, namely, that his "infinitely greater capacities" were not demeaned by a common ancestry with apes. He appealed to God to give him the strength "to be true to science" and expressed his hope that in science he and Huxley had the "common ground" on which to discuss "fairly and freely the deepest mysteries."[55]

Huxley began his response with thanks for the "hearty sympathy" and "frankness" that Kingsley had displayed and then turned to the reasons why he could not find consolation in the beliefs Kingsley espoused, despite their attractiveness. While having no a priori objection to the

[53] Kingsley to Huxley, 7 December 1859, HP: 19.160–1.
[54] Huxley to Dyster, 29 February 1860, HP: 15.110–12.
[55] Kingsley to Huxley, 21 September 1860, HP: 19.162–8.

immortality of the soul, Huxley nonetheless required that such mysteries be revealed to him in the same evidential terms as in an anatomical or physiological investigation. Lacking the "instinct of the persistence of that existence" that Kingsley and many others possessed, he could not see how accepting a belief on the basis of one's "highest aspirations" was anything other than "asking me to believe a thing because I like it." His own devotion to science taught him to guard against such preconceptions:

> Science seems to me to teach in the highest and strongest manner the great truth which is embodied in the Christian conception of entire surrender to the will of God. Sit down before fact as a little child, be prepared to give up every preconceived notion, follow humbly wherever and to whatever abysses nature leads, or you shall learn nothing. I have only begun to learn content and peace of mind since I have resolved at all risks to do this.[56]

As in his correspondence with his wife and his more orthodox friends, Huxley presented his scientific practice as an embodiment of the Christian virtues of humility, earnestness, and devotion. The common ground that he and Kingsley shared was based less on theories of development, modes of causation, or forms of explanation than on a code of credibility and honor involving the openness of belief, and the transparency of thought and language. Huxley confessed that he had spoken with more candor to Kingsley than to anyone except his wife. He admitted that he could not understand "the logic of yourself, Maurice, and the rest of your school," but that he would "swear by your truthfulness and sincerity." He had written frankly, he stated, because he hoped that through men such as Kingsley – men who were able to combine "the practice of the Church with the spirit of science" – the Church would not be

> shivered into fragments . . . an event . . . which will infallibly occur if men like Samuel [Wilberforce] of Oxford are to have the guidance of her destinies. Understand that this new school of the prophets is the only one that can work miracles, the only one that can constantly appeal to nature for evidence that it is right, and you will comprehend that it is of no use to try to barricade us with shovel hats and aprons, or to talk about our doctrines being "shocking."[57]

Huxley attacked the dismissive manners and comportment displayed in High Church circles toward the claims of men of science to truth. He suggested that someone who showed respect for his own modes of scientific conduct could assume the religious (even the miraculous) high

---

[56] Huxley to Kingsley, 23 September 1860, HP: 19.169–76.
[57] Huxley to Kingsley, 23 September 1860, in L. Huxley ed. 1900, 1: 217–22.

ground. Huxley's pledge of support for an Anglican Church, filled with
the holy spirit of science and presided over by clerics like Kingsley rather
than Bishop Wilberforce, came just a few months after Wilberforce had
reviewed the *Origin of Species* unfavorably in the *Quarterly Review* and at
the British Association. At this time, the bishop was leading a campaign
within the Church against the authors of *Essays and Reviews*, as well
as a movement, widely opposed by men of science and liberal clergymen, for public prayers to assist in the relief of national calamities like
drought and cattle plague.[58] In October 1860, Huxley referred to a plan
to write an article jointly with Kingsley on weather prayers for *Fraser's
Magazine* and to the "duty of all men of science" to aid Kingsley in this
matter. He and Kingsley exchanged letters during March 1861 regarding
a projected memorial on matters of belief in support of the authors of
Essays and Reviews to be signed by "scientific men." Kingsley stated that
such a testimonial would be "most valuable."[59] Wilberforce's opposition to *Essays and Reviews*, and his campaign to redress natural disasters
with public prayers, subordinated science to a theological interpretation
of the Bible and of natural events. Kingsley's support for the testimonial signed by scientific men, however, seemed to endorse the power of
scientific method as an instrument for understanding not just natural
phenomena, but Scripture; it also implied that men of science were the
equals of clergymen in the interpretation of nature and the Bible.

The precise nature of the support that Kingsley drew from Huxley
and from science was a matter of extensive private debate, however. In
October 1860, Kingsley remarked that he was reading Spencer's *First
Principles* of synthetic philosophy with "respect and care," although he
thought it hopeless to "apply exact science to, or to find inevitable laws
in, the history of mankind." Huxley responded in the fashion of one
of his addresses to an uninstructed public: "scientific method" was not
"a method of inquiry but the method and the necessary method of all
inquiry." Human history had manifested no other. Religions, including
Kingsley's own Anglicanism, were true to the degree that they were
based on sound "moral inductions." Kingsley's remarks drew upon the
writings of liberal theologians, like *Essays and Reviews*, which employed
scientific methods in the investigation of God's will in history. As a
clergyman, he had a duty to interpret natural laws as expressions of
Divine will. For Huxley, however, this clerical practice of science was
self-defeating; for the proper use of scientific methods of investigation

---

[58] See Turner 1993: 151–70.
[59] Huxley to Kingsley, 4 October 1860, HP: 19.191; Kingsley to Huxley, 16 October 1860,
HP: 19.193–4, and 12 March 1861, HP: 19. 202. For a discussion of this projected memorial, see Brock and MacLeod 1976.

entailed the adoption of materialist modes of explanation in which Divine (or human) will had no place. He would argue many times in public that this use of materialist language involved no commitment to materialism – that it involved no assumptions about the ultimate substances or powers underlying phenomena. For Kingsley, however, the application of such materialist terminology to the interpretation of human events would eliminate Divine agency from the course of history. According to Kingsley, human beings, because they possessed a soul, had the ability to command natural laws and to master nature. Man, by virtue of his special nature, could transcend the bounds of the natural laws that Huxley was continually invoking. Like other liberal theologians, Kingsley thus appealed to a version of science with inherent limitations. By contrast, Huxley repeatedly attacked the notion that natural laws were manifestations of divine (or human) agency. Such claims tipped the balance of power over the interpretation of nature in favor of clergymen.

During the next several years, the two men corresponded about the best way of making science religious, and religion scientific. In July 1862, Kingsley explained to Huxley that Darwin's theories had opened "grand views" and a "new era" to him, and that from these hints – "for they are no more than hints" – all natural theology would be rewritten and "a whole new science" arise.[60] In December of the same year, Kingsley sent Huxley a sketch of this new science and theology in the form of a children's story, illustrating that the wisest God was one who "makes all things make themselves."[61] Huxley responded with a copy of *Man's Place in Nature*. Diminishing the importance of "Colenso controversies" and other attempts at critical revisionism of the Bible, Huxley pointed to "the impassable gulf between the anthropomorphism (however refined) of theology and the passionless impersonality of the unknown and unknowable which science shows everywhere underlying the thin veil of phenomena."[62] Kingsley's new science of natural theology discerned evidence for a personal God, Holy Spirit, and individual soul. It served to redeem a Malthusian nature in which humans were beasts and to reinforce the traditional paternal duties of the clergyman as a source of strength and consolation in the parish, as a commanding interpreter of human history, and as a presider at natural events like birth and death. In denying that nature afforded evidence for a personal God, Huxley was undermining a crucial foundation for this pastoral role, while he was making room for the more impersonal ministrations of men of science.

[60] Kingsley to Huxley, 18 July 1862, HP: 19.205–6.
[61] Kingsley to Huxley, 20 December 1862, HP: 19.211.
[62] Huxley to Kingsley, 30 April 1863, HP: 19.215.

In a letter of May 1863, Kingsley tried to clarify that whereas Huxley based the distinction between humans and apes on brain function, he, Kingsley, located the distinction in the "soul": "What will you do with a fellow who persists in thinking such stuff as that, knowing it is sounder science than yours?"[63] Huxley reponded: "If you tell me that an Ape differs from a Man because the latter has a soul and the ape has not, I can only say it may be so; but I should uncommonly like to know how either that the ape has not one or that the man has."[64] Huxley's disagreements with Kingsley turned on the forms of evidence permissible in establishing matters of fact, on the role of the will in evaluating evidence, and on the nature of human agency and its relationship to material causes and laws. These differences, crucial in negotiations over cultural authority between men of science and the clergy, were not suppressed in their correspondence; indeed, they formed the basis of a protracted and friendly discussion. At the junctures where their discussion broke off, each fell back upon shared assumptions about the limits of knowledge, assumptions that had framed their relationship from the start. "As long as you allow our infinite ignorance of what may be," Kingsley wrote, "we shall get on together as jolly as two turtledoves." Huxley replied, "If any expression of ignorance on my part will bring us nearer we are likely to come into absolute contact, for the possibilities of 'may be' are, to me, infinite."

By confessing their ignorance regarding metaphysical matters, and by producing a shared ethics of belief itself, Huxley and Kingsley built a common ground on which they could overcome the enormous disparities in their practices. Agnosticism, as Huxley came to define it, was certainly averse to the preeminence allotted in matters of belief to faith and authority (whether scriptural or institutional) by many religious organizations, Anglican, Roman Catholic, and Nonconformist. Yet the principles of belief were themselves undergoing revision, especially among liberal Anglicans. By Huxley's own account, agnosticism was invented in the 1870s, amid his debates with other high-minded cultural authorities in the Metaphysical Society. But Huxley and Kingsley were already practicing agnosticism in their correspondence in the early 1860s. In effect, such methodological imperatives were forms of sociability, ways of conducting oneself with others whose beliefs were profoundly different. In private, "you and I," as Kingsley put it, "have hold of exactly opposite ends of the string and yet are feeling on till we meet in the middle."[65]

---

[63] Kingsley to Huxley, 14 May 1863, HP: 19.221–2.
[64] Huxley to Kingsley, 22 May 1863, in L. Huxley ed. 1900, 1: 242–4.
[65] Kingsley to Huxley, 14 May 63, HP: 19.221–2.

In May 1863, Huxley wrote supporting the argument of an unspeci-
fied manuscript by Kingsley that criticized those who maintained an impassable "intellectual gulf" between man and the lower animals. Huxley
interpreted Kingsley's reference to the use of a particular phrase by
Charles Lyell as an application of Darwin's views of human history:

> The advance of mankind has everywhere depended on the production
> of men of genius; and that production is a case of "spontaneous varia-
> tion" becoming hereditary, not by physical propagation, but by the help
> of language, letters and the printing press. Newton was to all intents and
> purposes a "sport" of dull agricultural stock, and his intellectual powers
> are to a certain extent propagated by the grafting of the "Principia," his
> brain shoot, on us.[66]

Huxley here conflated human agency and culture with forces of nature.
Human culture was what distinguished man from the lower animals;
both Huxley and Kingsley, as the makers of language, letters, and books,
could agree on this preeminent role, even if Huxley naturalized culture.
The production of this culture involved a reconstruction of gentlemanly
credibility through new discourses and practices of learned behavior,
especially those of belief. Through displays of openness, aversion to
authority, breadth of knowledge, and depth of religious feeling, men
of science and liberal clerics equipped themselves as moral leaders and
attacked the manners that had distinguished a previous generation of
learned gentlemen.

Kingsley and Huxley shared an identity as social leaders, interpreters
of nature, and producers of culture. Yet this identity was not sustained
only by the processes of mutual definition and appropriation that characterized their correspondence. Their status as elites required that they
be acknowledged as scientific and religious authorities by others. A
whole range of social groups were important in the conferring of cultural authority, both in the public settings where they functioned as audiences and in the more private contexts of friendship, family, and work
and patronage relations. Through moral discourses about method; the
Christian virtues of inquiry and conscientious doubt; and the codes of
honesty, transparency, and plain speaking governing the formulation
of belief, men of science and religion conducted themselves critically,
candidly, and disinterestedly before audiences who might otherwise
question the value of liberal culture – especially biblical criticism and
evolution by natural selection – and regard their discourse as factional
and uncertain.

The reformation of science and religion was thus not only a matter of
self-fashioning and self-presentation by elites; it was also a process of

---

[66] Huxley to Kingsley, 5 May 1863, in L. Huxley ed. 1900, 1: 240–2.

reforging social bonds on a national scale, of refashioning a public. In 1871, Huxley served as a member of the London School Board, which was the first of many local councils set up during the previous year by state legislation to prepare a program for England's first national system of elementary schools. Together with the clergymen who sat on the committee that designed the curriculum for London's schools, Huxley presided over the process in which scientific and religious subjects were first introduced systematically to English schoolchildren. How did men of science and clergymen make the classroom into the antechamber of their broad church?

The Classroom

In 1870, the House of Commons, under the leadership of the Liberal M.P. William Forster, passed a bill committing public funds to the building of new elementary schools and calling for the election of local boards to procure land and materials, prescribe curricula, and appoint teachers for these schools, as well as to oversee the many existing learning institutions receiving government aid. In some respects, this legislation was part of a process of state intervention in education that had begun in the 1830s with the establishment of grants to private schools that favored certain textbooks and examination subjects and with the creation of a council and inspectorate for education. By the time of the 1870 Act, the government had committed public money toward the building of over five thousand new schools and the improvement of twenty-three hundred others, and was receiving roughly three thousand applications for aid every six months.[67]

Under the largely private and voluntary system of education thus established in Britain, however, most poor children received either no schooling or attended school only part-time – several hours a day, or on Sunday – until age ten or twelve, when they began full-time employment. Promoters of the 1870 measure linked it quite explicitly to recent changes in the suffrage: the 1867 Reform Bill had enfranchised many members of the urban working classes, making them the majority in most boroughs. Speaking before the House of Commons on 17 February 1870, Forster stated that the new legislation was directed toward those whom previous measures had left unhelped, "the children of the working classes." For this segment of the public, which he saw as incapable of supplying or even demanding what it needed most, state intervention was urgently required: "Upon this speedy provision

[67] On reforms in elementary education for the working classes, see Simon 1960 and 1965 and Layton 1973.

[of elementary education] depends ... I fully believe, the good, the safe working of our constitutional system."[68] Such policies complemented a variety of institutions – museums, gardens, libraries – that had been recently built or refashioned for the edification and healthy recreation of workers and their families in the cities. In part, elementary education was to be a means of creating and sustaining the sort of audiences to whom Huxley pitched his Friday evening and Sunday afternoon lectures on natural history: the "large masses" who no longer regularly attended places of worship.

Since the process of state intervention in education had begun, advocates of secular instruction, like the phrenologist George Combe, had often argued that the denominational jealousies of religious bodies were the chief impediment to the establishment of a national education system. The enduring rivalries between denominations were also cited by many proponents of scientific education, like the editor of *Nature*, Norman Lockyer, as indicative of the divisiveness of theology, and of the failure of religious institutions to provide a universal social bond.[69] The 1870 Act did not make education secular. It assumed that schools were places of religious practice and instruction; however, it tried to regulate religious education in a way that did not expose the state to charges of sectarianism. Prior to the Act, virtually all schools and teacher-training institutions had been connected with a specific religious denomination. The 1870 bill kept religious affiliations private by eliminating denominational inspection (including inspection on religious subjects), by stipulating that schools receiving grants should not pronounce upon attendance at a particular Sunday school or place of worship, and by creating conditions in which parents could withdraw their children from any "religious observance" if they wished.[70] To overcome denominational divisions, promoters of the bill frequently evoked Scripture as a common denominator. Speaking at the Birkbeck Institution in 1870, Forster stated, "I have the fullest confidence that in the reading and explaining of the Bible, what the children will be taught will be the great truths of Christian life and conduct, which all of us desire they should know, and that no effort will be made to cram into their poor little minds, theological dogmas which their tender age prevents them from understanding."[71] In election campaigns for the London School Board, virtually all the candidates promised to be nonsectarian and to give

---

[68] Hansard's *Parliamentary Debates, 1870*, vol. 199, pp. 443–4.

[69] On secular education, see, for example, Combe 1852 and Lockyer 1906.

[70] For these religious ordinances, see Article 7 of the Education Act in *Parliamentary Papers, 1871*, vol. 21, Appendix: xxii.

[71] Forster, speech at the Birkbeck Institution, as reported in *The Times*, 7 June 1870.

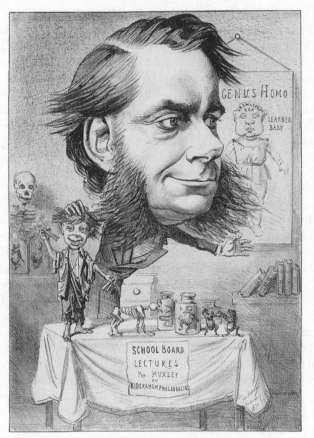

Plate 7. Huxley on the London School Board, lectures on
"Kidcramumphulononsins" (*The Hornet*, 1 March 1871.
Courtesy of the Wellcome Library, London).

the Bible central importance. State schools would thus further a common Christian culture by featuring the Bible in the curriculum. Recently
endowed with political agency, and without the customary superinten-
dence of a parish priest, a new generation of urban workers were to
be taught the principles of religious life at school. The new classrooms
were to be their surrogate church.

Huxley was one of forty-nine candidates returned at the November
elections for the London School Board. He had stood on a platform
of "scientific education" and had been endorsed by the *Times* as the
representative of "men of science." He finished a distant second in
the district of Marylebone to Elizabeth Garrett Anderson, then se-
nior physician at the dispensary for women and children on Euston

Road. Though unable to vote, women of the upper and middle ranks were widely respected as educators of children, and they possessed authority – irrespective of their gender – over workers. Among the other School Board members were five M.P.s and eleven clergymen from a variety of denominations, most of whom were distinguished by some involvement with educational institutions or children's philanthropy. Working men had stood as candidates in all boroughs, but none were returned. A month prior to the election, a deputation of workers had unsuccessfully petitioned the Privy Council to extend the early closing hours of the polls, which required many workers to leave their jobs to vote.[72] Forster rejected the petition, citing the inconvenience of counting votes. He added that this year was "an experiment" and that any faults would be corrected next year. Nor were workers among the hundreds of witnesses summoned by the London board, and whose experiences counted as evidence for the board's decisions. The absence of workers was not incidental to the operations of the board, which was enormously divided by religious denomination, profession, and gender. Huxley, for example, had spoken on many occasions of "parsondom" as the enemy of science, and of women, who were prone to superstition and ruled by their feelings, as the natural allies of parsons.[73] The differences among the elites who made up the London board would be overcome, however, because, as educators, they shared a relationship with the newly enfranchised, and excluded, working classes.

Witness reports collected by the London board in the early months of 1871 testified both to the heavy reliance of most elementary educators on the Bible and to the variety of uses, from moral and theological instruction to the rudiments of reading and counting, to which the book was put.[74] Severe disagreements among board members over the status of the Bible in the classroom were aired in a series of debates during February and March. In most instances, the clergymen, together with Huxley, proposed legislation. Opening the discussion on 8 March, Reverend Joseph Angus, a biblical scholar and president of the Baptist College at Regent's Park, maintained that morality could not be taught without the Bible; Reverend James Picton, a Congregational minister in Hackney, responded that the conscience, being divinely taught, required no book and was best cultivated at church or in Sunday school. Speaking against Picton were John MacGregor ("Rob Roy"), who insisted that Christianity must be taught, and Sir Charles Reed, board chairman and

---

72 On the London School Board elections, see *The Times* 15, 20, 29 October, 29 November, and 1 December 1870.

73 See especially Huxley 1865 on the "natural alliance" of parsondom and women and its effect as a "drag on civilization."

74 *First Report of the Scheme of Education Committee*, pp. 11–24.

M.P. for St. Ives, who remarked that outside of board schools, many children received no moral training. A special provision for translations of the Bible approved by Roman Catholics was debated and rejected. Huxley motioned unsuccessfully that the Bible be taught as a part of "religious culture."[75]

The 8 March debate was an attempt to overcome a deadlock that had ended the session several weeks earlier. On 15 February, it had been proposed "that in the Schools provided by the Board the Bible shall be read and instruction in religious subjects shall be given therefrom." But this had been countered by another motion to bar Bible reading and religious instruction entirely as being opposed to the "conscientious convictions of many Ratepayers and Parents, and as leading ultimately to a denominational system of Religious teaching in the schools."[76] Tight control of the legislation process was threatened when newspaper reports of the proceedings prompted members of the local electorate to enforce their own interpretation of the Education Act in the face of school board factionalism. At the next meeting, the board received a petition signed by 930 voters, stating that if Bible "reading" was to be unsectarian, it could not be combined with religious "instruction."[77]

When the board members reconvened in March, they were thus assembling not only to resolve their internal differences but also to defend their status as leaders and educators who were authorized to interpret sacred texts and furnish religious instruction to the public. By calling for Bible reading independent of religious teaching, the ratepayers had pointed to the failure of any existing group to represent religion in a way that was perceived as universal. Their petition suggested that the new working classes would respect only a united religious authority, not a schismatic one. In reaching agreement over the place of the Bible in the classroom, board members thus were not merely creating the basis for a common Christian culture, but were constructing themselves as a recognizable elite transcending the fragmented denominational societies, councils, and churches (as well as the factional politics) presently controlling education.

Just how the London School Board members were able to reconcile their differences is suggested by Huxley's involvement in the proceedings. In his speech before the board on 8 March, Huxley made a plea for

---

[75] *London School Board Report*, minutes for 8 March 1871, p. 81; *The Times*, 9 March 1871.

[76] The motion against the use of the Bible was made by T. Chatfield Clarke. *London School Board Report*, minutes for 15 February 1871, pp. 62–3. "Ratepayers" corresponded to voters, which after 1867 included most urban working men. Only the poorest parents would not have qualified to vote and would not have been assessed the local taxes to support board schools.

[77] *London School Board Report*, minutes for 1 March 1871, pp. 75–6.

science as a party to negotiations over religious instruction. He remarked that the religious difficulty was not just between the old Catholic Church and a splintered Protestantism, but that there was another side to these matters – the "scientific aspect" – which had a religion and morality of its own. He added, however, that the great mass of "low half-instructed population" owed what little redemption from ignorance and barbarism it had mainly to the efforts of the clergy and that any system appealing to these people must be connected with their own system of belief.[78] His position was an abbreviation of one he had outlined in an article that the editor of the *Contemporary Review* had released to the newspapers shortly before the school board election:

> I have always been strongly in favour of secular education, in the sense of education without theology; but I must confess I have been no less seriously perplexed to know by what practical measures the religious feeling, which is the essential basis of conduct, was to be kept up, in the present utterly chaotic state of opinion on these matters, without the use of the Bible.[79]

In this platform, Huxley proposed that the passions men of science had dismissed in their debates with clergymen as foundations for belief and action be rehabilitated for the instruction of the working classes and, further, that the text that had formed the subject of endless criticism and qualification in these debates be raised up as a moral foundation. When confronted by low and unlearned masses armed with the vote, the differences that divided learned elites, and that formed the very substance of their discussions and polite society, counted for little. Huxley could speak favorably of religious feeling in the children of workers, for such feelings were the appropriate bases of action for classes (as well as races and sexes) lower on the developmental scale.[80] For these classes, the psychological and sociological theories of religion produced by men of science and those of liberal theologians were in agreement.[81] By presenting a view of conduct based on religious feelings, and by supporting a text full of narratives to arouse such feelings, Huxley underlined the need for authorities whose tasks were to insure that the feelings inspired were the correct ones and to indicate the reasons and laws that supported them.

---

[78] Huxley, speech before the London School Board, as reported in *The Times*, 9 March 1871.

[79] Huxley 1870b: 154.

[80] On the ranking of sexes and races in Victorian psychology and anthropology, see, for example, E. Richards 1983, Lorimer 1988, and Russett 1989.

[81] For accounts by men of science and liberal clergymen that based religion in emotions such as dependence, awe, and reverence, see Maurice 1837, Mansel 1858, Tyndall 1874, and Temple 1884.

At the end of the 9 March session, the board approved a modified version of the original motion regarding Bible reading and teaching by a vote of thirty-eight to three. Huxley was in favor of the measure, which established "that in the Schools provided by the Board the Bible shall be read and there shall be such explanations and such instruction therefrom in the principles of morality and religion, as are suited to the capacities of children."[82] The edict did not prescribe a uniform interpretation of the Bible. It did, however, dictate a uniform function for the Bible: the teaching of universal morality. It also imposed a uniformity of method and practice: the appeal to principles, and the use of explanations and instructions that ruled out the possibility of literal readings. These were the same methods that contributed to the common ground for men of science and theologians – liberal and conservative – as constructed, for example, in their debates on the formation of belief. These practices constituted the basis of the authority of such groups over nature and Scripture, and over the more local and private activities of reading and interpretation. By rewording the edict, the board members reinforced the authority of the Bible as a source of universal truth. They also undermined the authority of local voters, the working classes in particular, to read the Bible for themselves. In effect, they reasserted the board's identity as a body able to transcend denominational differences and represent a common Christian culture precisely at the moment when the local press had shown that its members were not unified but divided on issues of religious truth. By mandating educators to teach the Bible as the embodiment of this common culture in the classroom, the board both completed and concealed its work in settling these divisions.

The solidarity displayed by the board members on 9 March was soon tested again, however. Several weeks later Huxley proposed "that in all elementary schools in which the Bible is read a selection from the Bible, which shall have been submitted to and approved by the Board, shall be used for that purpose." Such selections, he stated, would be based on both the "moral aspect" and the "scientific aspect" of the text; for while certain stories, like that of Lot, were ethically unfit for children, others, like the seven days of creation, contained statements "that men of science demur." He added that the board could not leave such matters "to the chance wisdom" of the people. Huxley's motion underscored the alliance that had been struck by scientific and religious elites in the constitution of authority over working-class children. But it also called for a shift in the ostensible locus of power from the biblical text to the board itself. The motion was impolitic, for it implied that religious authority resided in learned elites rather in the revered source of which they were

---

[82] *London School Board Report*, minutes for 8 March 1871, p. 81; *The Times*, 9 March 1871.

the proper interpreters. To ban portions of the Bible as unscientific or immoral was to reintroduce divisions and qualifications where there was supposed to be unity. It was to risk making explicit the same problems of evidence and belief that divided learned groups in, for example, debates over evolution. Objecting to Huxley's proposal, the Reverend Anthony Thorold urged that such a measure would dishonor the Book and the teachers who edited it. The motion was defeated thirty-five to five.[83]

It was in this kind of classroom, where the Bible was the universal source of truth and the teacher the universal medium of religious instruction, that science was first introduced systematically to elementary school children. Prior to the design of the curriculum, the London School Board had gathered testimonies from local schoolmasters that indicated that elementary courses in physiology, physics, natural history, and social science were already widely taught by 1871, except at schools in the poorest districts.[84] Precisely how these subjects were taught and to what end was less clear. The 1871 Report of the Science and Art Department – the government bureau that controlled the content of such courses indirectly through its examination system – identified scientific instruction quite explicitly with commercial and industrial use. Many of the courses administered by the department were sponsored by local industries and designed to serve the needs of those industries.[85]

On 8 February, Huxley made a speech before the board, proposing the appointment of a "Scheme of Education Committee," with himself as chair, to work out the details of the elementary curriculum and present its conclusions before the assembly for approval. Objections were raised that the proposal was premature and that such matters were best determined by the board as a whole. In the *Times* report of Huxley's speech, he was said to have stated the chief aim of the school curriculum to be "the implanting in the minds of children and giving them reasons for the great laws of conduct in this world, and the primary one of religion and morality." "All the laws of conduct," he added, "and all the great principles of morality might be taught a child apart from disputed points."[86] While the board was settling the status of the Bible, Huxley presented himself as someone who would ensure that other subjects in the school curriculum would complement the place of the Bible by providing the reasoned interpretations for the feelings that biblical narratives were supposed to evoke. Elected on a platform of "scientific

---

83 *Times*, 30 March 1871.
84 *First Report of the Scheme of Education Committee*, pp. 11–48.
85 *Parliamentary Papers, 1871*, vol. 24, pp. xxviii–xxix.
86 Huxley, speech before the London School Board, as reported in *The Times*, 9 February 1871.

education," Huxley indicated that the role of science in education would be to demonstrate that nature was a place of order and hierarchy and that social order and social hierarchy were natural. The *Times* reporter paraphrased the conclusion of his speech, which supported the existing stratification of primary schools – ragged or free schools for the poorest, evening and half-time schools for those compelled to work, day schools for the rest – a conclusion that evoked great applause:

> He believed that no education system in this country would be worthy of the name of a national system, or fulfill the great objects of education, unless it was one which established a great education ladder, the bottom of which would be in the gutter and the top in the University, and by which every child who had the strength to climb might, by using that strength, reach the place for which nature intended him. (loud cheers)[87]

Huxley presented the school board with a list of nominees for his committee, which included four clerics, one woman, and "every shade of opinion" on the board as a whole. His motion was seconded by Reverend Joseph Angus, who added that after such a speech, the necessity of a Scheme of Education Committee could not be doubted. The motion was passed unanimously. Citing from their own gentlemanly code, two of the clerics whom Huxley had nominated for the committee, Dr. Alfred Barry, a liberal theologian and principal of King's College, London, and Benjamin Waugh, a philanthropist and Congregational minister at Greenwich, later praised the "singular candour" with which Huxley had conducted himself on the board, his "commonsense," his "lofty ideals," his "knowledge of every subject connected with culture," and his "honesty of a child."[88]

In the curriculum that Huxley and other committee members designed, "morality and religion" as embodied in the Bible received first place, followed by the rudiments of reading, writing, and arithmetic; social economy; drawing; domestic economy for girls; and an elementary course in physical science.[89] To advance the last, Huxley produced his own sacred text in 1877, *Physiography: An Introduction to the Study of Nature*. Promising to lead the pupil toward a clear picture of the principles pervading the natural world, Huxley proceeded to fill the Thames basin with order, summoning images of the earth's laws and history from the smallest pebble in a child's parish. Evolutionary ideas were introduced tacitly, through discussions of an extended time scale and the uniformity of causes.

[87] Huxley, speech before the London School Board, as reported in *The Times*, 9 February, 1871.
[88] L. Huxley ed. 1900, 1: 351–3.
[89] *London School Board Report*, pp. 158–9.

Conclusion: Metaphysical Society behind Closed Doors

Religious symbols and discourses remained fundamental to the possession and exercise of authority for Victorians.[90] They were also important in the sealing of domestic bonds and the maintenance of enduring friendships. Huxley did not adopt a religious persona merely for strategic or satirical purposes; nor was his a persona often designed to rock the foundations of the Anglican establishment. If he had not incorporated religious forms in his manner, style, terms of address, and epistemology, he would have had few patrons, tiny audiences, and a poor livelihood; he would also have had few friends and perhaps no family. Religion helped to form the basis of scientific identity in the second half of the nineteenth century through a process that involved a variety of public and private negotiations with family members, with friends, and perhaps above all, with clergymen, who drew in their turn on scientific discourses and practices to reshape their own vocation.

Through the use of biblical narratives and heroic histories of science, cultured men reworked their prophetic and priestly personae for audiences they presumed would always be as children, in need of wise counsel on matters of belief. Such storytelling was not purely pedagogic. It was a favored form of sociability in exclusive learned societies. In 1885, R. H. Hutton, the liberal cleric and editor of the *Spectator*, wrote a reminiscence of the Metaphysical Society, which he had helped to found in 1868, together with the Poet Laureate, Alfred, Lord Tennyson, and the Reverend Charles Pritchard, Savilian Professor of Astronomy at Oxford. In his piece, Hutton reconstructed one of the monthly soirees, in which clerical figures like Stanley, Martineau, and the Roman Catholics Henry Manning and Wilfrid Ward met for dinner with men of science such as Huxley, Tyndall, and John Lubbock to discuss religious questions "after the manner and with the freedom of an ordinary scientific society."[91] The topic of the evening was which among the learned groups present used the expression "I believe" with more solemnity. Debate commenced when Dr. Ward, who wore the "stamp of definite spiritual authority," described how the moral conditions of human nature dictated the belief in God. Huxley followed, "his slender definite creed" concealing "the cravings of his large nature," stressing that men of science were humbly bound to require firmer grounds for belief than the testimony of personal feelings. Father Dalgairns replied that theologians did not rest their

---

[90] On the enduring authority of Christianity in nineteenth-century Britain, see Gilbert 1976, Obelkevich 1976, McLeod 1984, and Hilton 1988. On the relatively marginal status of secularist movements, see Mullen 1987.

[91] Hutton 1885: 177–96. There were 59 original members, 16 of whom were clerics. On the Metaphysical Society, see Brown 1947.

deepest creeds on a "working hypothesis," but on a "higher intuition than any inductive law can engender." Conversation shifted to assorted reflections on human and miraculous agency. John Ruskin remarked on the inconstancy of man's noblest powers. Walter Bagehot dwelt upon men's natural credulity and longing for marvelous displays, which overwhelmed the hard discipline produced by disappointment, especially in "the City, where hopes are crushed every day."

In Hutton's account, the learned members of the club conducted themselves as England's spiritual masters and wisest counselors. They presented the highest intuitions, the sternest discipline, and even the miraculous to a population in urgent need of transcendence: "Everything ... spoke of the extraordinary fermentation of opinion in the society around us." In their learned papers and conversations, such men seemed to divide the world among themselves, to pose and resolve the problems of society as a whole in their debates with one another. The Metaphysical Society was a religious body representing different denominations, including Science; it also offered the possibility of a universal religion that transcended sectarian divisions through its members' ability to settle disputes in a gentlemanly manner. By this means, they could claim to embody in a gentlemanly society those beliefs and values that were of fundamental importance to everyone, paralleling contemporary debates among the Whigs and Tories in Parliament over which party best represented the "common man."[92]

Hutton wrote his piece nearly five years after the Metaphysical Society closed its doors. As one broadsheet produced in the early 1880s indicates, the claims of learned men, including members of Parliament, to be the best representatives of England's people could become the subject of derision (see Plate 8).[93] As the Church parties pull in different directions beneath the dome, Roman Catholics, Dissenters, Freethinkers, and Secularists stake out different terrain outside. In the upper left corner, John Tyndall and Herbert Spencer accompany Huxley toward the dawn of Darwinism and Protoplasm (?). Among the liberal Anglicans featured are Frederick Farrar ("who really wants a hell when he can enjoy this?") and the late J. W. Colenso, following the scientific path. A party of Dissenters tugs at the foundations of the established Church. Moncure Conway, founder of the Association of Liberal Thinkers, beckons toward the scientific horizon. Although including a number of former members of Hutton's Metaphysical Society, the leaders of this "National Church" do not appear as earnest,

---

[92] On the language of representation used by Liberals and Conservatives in debates over the enfranchisement of the working classes, see Cowling 1967: 48–60.

[93] A different version of this broadsheet, printed ten years earlier, is reproduced in Desmond 1998.

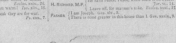

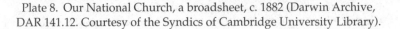

Plate 8. Our National Church, a broadsheet, c. 1882 (Darwin Archive, DAR 141.12. Courtesy of the Syndics of Cambridge University Library).

Plate 8. (*continued*)

conversable gentlemen; whether as rowdy showmen or as misguided mountaineers, these small bands of learned men have no visible connection to one another. Science, it is suggested, is perhaps just one faction among many. This depiction clearly undermined the claim, advanced in the Metaphysical Society, that it was possible to achieve consensus through learned discussion. Among the drumbeaters crying their creeds in the foreground of the picture is a "pious mountebank": William Booth, founder of the Salvation Army, who would clash with Huxley during the 1890s for the right to represent a group not visible anywhere in the picture – the working class.

5

"Darkest England"

Science and Labor in the 1880s and 1890s

In tens of thousands of poor homes, Huxley's name must be [that of] one who was the mere tool and instrument of a prevalent orthodoxy – a despotism personified.

– Daylight, 13 July 1895

Chapters 3 and 4 have shown how Huxley drew on literary and religious models to shape scientific identity, and how he contended and cooperated with men of letters and clergymen in the reform of curricula in public schools and universities, and in new elementary classrooms supported and administered by the state. The public controversies in which Huxley engaged with other leading social figures took place before a variety of audiences who often had quite different notions of what science, literature, and religion should be, and who did not merely defer to learned men. Thus the meaning of the cultural practices in which such men engaged, the values that they attached to their work, and their self-definitions were in large part dependent on the views of the people whom they sought to educate. The "cultural authority" for which learned groups contended was itself in question and could be acquired only by meeting the expectations of various publics. Men of science

135

were called upon by their patrons in government and industry to be meritocratic and practical, while many of their audiences and publishers expected rational amusement, moral didacticism, and a display of specially endowed powers of intellect.

From the beginning of his tenure at the School of Mines through the 1870s, one of the publics with whom Huxley was most popular was "working men." The expression most commonly referred to artisans, but it was also very often a moral designation, signifying the Victorian values of industriousness and individual improvement. Working men were the principal audience for Huxley's extremely successful evening lectures at Jermyn Street. Speaking before such an audience, Huxley placed the sciences firmly within the liberal program of material progress and social amelioration through education. In accordance with this program, Huxley could criticize abuses and signal inequities in the capitalist system by encouraging workers to better themselves through learning and by admonishing owners for hindering their efforts to do so.[1] He was not expected to advocate fundamental changes in the system of government, the structure of private enterprise, or the distribution of wealth.

In the 1880s, the liberal values and assumptions that underpinned Huxley's role as an an educator and improver of workers were severely challenged. Strong workers' movements – new unions, learning societies, a new political party – began to redefine the dignity of labor and the sort of practices deemed productive and good.[2] Also contested were elite models of leadership in which ability was attached to property and authority was based on the discipline, and even the denigration, of the passions. Huxley, who had long presented himself as a selfless and nonpartisan man of science, came to be ranked by some as among the brutally selfish ruling classes. To preserve his identity as a social reformer and moral authority, he set about reinterpreting the conditions that had alienated so many, some men of science included, from the established order. To do so, he would denounce alternatives to this order, and actively defend the foundations of a commercial and industrial system about which he had always been ambivalent. The very process of defending himself, however, would cut to the core of his identity as man of science.

On 1 December 1890, Huxley entered into one of his last public controversies, writing in a letter to the *Times*:

---

[1] For accounts of Huxley's lectures to working men, see Desmond 1998: 208–11, 292–5.

[2] For classic discussions of the formation of the English working class, see E. P. Thompson 1963, Gray 1981, and Hobsbawm 1984. For revisionist approaches to the history of "class" see, for example, G. S. Jones 1983, Joyce 1991, and Warhman 1995.

Few social evils are of greater magnitude than uninstructed and unchastened religious fanaticism; no personal habit more surely degrades the conscience and the intellect than blind and unhesitating obedience to unlimited authority. Undoubtedly, harlotry and intemperance are sore evils, and starvation is hard to bear, or even to know of; but the prostitution of the mind, the soddening of the conscience, the dwarfing of manhood are worse calamities. It is a greater evil to have the intellect of a nation put down by organised fanaticism; to see its political and industrial affairs at the mercy of a despot whose chief thought is to make that fanaticism prevail; to watch the degradation of men, who should feel themselves individually responsible for their own and their country's fates, to mere brute instruments, ready to the hand of a master for any use to which he may put them.[3]

The organized fanaticism to which Huxley refered was the Salvation Army; the master and despot was William Booth, its founder and general. What occasioned Huxley's ire was a book Booth had published several months earlier, *In Darkest England and the Way Out*.[4] In part a graphic account of the conditions of the urban poor, in part a scheme of improvement, the book outlined a system of workshops, cooperative farms, and labor colonies designed to raise England's sunken population from despair. Booth was an uncharacteristic antagonist for Huxley. He was not a man of letters or a high churchman, but a former Methodist preacher, now an associate of labor leaders and a popular figure among considerable numbers of the urban poor.[5] Huxley had often used the *Times* for occasional correspondence, but the Booth affair elicited twelve letters extending over several months. In bitterness and vehemence, Huxley's attacks on the general and the blind enthusiasm of the Army surpassed even his own previous invective. Booth's agenda threatened Huxley's own in a way that more traditional rivals, namely, other learned groups like Anglican bishops and university dons, did not. Booth's *Darkest England* scheme called into question beliefs about "working men" and forms of leadership and social organization long taken for granted by Huxley and other men of science, and upon which their authority as learned men rested. The *Times* debate over *Darkest England* was only one of a series of public controversies in which Huxley engaged in the 1880s and 1890s. In these disputes over land nationalization, technical education, Irish Home Rule, and the authenticity of miracles, he and other men of science worked to solve the latest "condition of England" question and to ensure that the question was discussed on their terms.

[3] *Times*, 1 December 1890. These letters from December 1890 and January 1891 were published with supplementary material in Huxley 1891a.
[4] W. Booth 1890.
[5] On Booth and the Salvation Army, see Ervine 1934, Collier 1965, and Sandall 1955.

"A Copious Shuffler"

In May 1885, under the pressure of serious illness, Huxley resigned his professorship at the Royal School of Mines and, within six months, his presidency of the Royal Society. He remained a member of the governing body of Eton school, the Royal Society council, and the London University Senate, but he was seldom present at their meetings. Withdrawal from such positions of distinction and administrative power did not come easily. To his friends he had to excuse himself repeatedly for inactivity, and to insist that science itself had never been a burden, only the public responsibilities it incurred. In his last presidential address to the Royal Society, he assured his scientific fellows that he would not lapse into idleness, but take up more anchoritic labors. To heads of state he described his retirement as "peace with honor."[6] The expression was also a plea. How was Huxley to maintain honor now that the positions that conferred it had been left behind? As the Treasury Department was deciding the terms of his pension, he was compelled by codes of scientific conduct to be indifferent to monetary figures. He had also spoken out on occasion against one customary mode of rewarding scientific achievement: titles of nobility. Newton and Cuvier, he said, had "lowered themselves" by accepting knighthoods.[7] But indifference to money was also an aristocratic tradition. It may have helped to gloss over some uncomfortable truths: that Huxley needed a large settlement if he was to live comfortably, and that a substantial cash sum was the only way for an enlightened state to confirm the dignity of thirty years of scientific work that had shattered his health.

Huxley was awarded his full salary plus a Civil List pension (£1,500 a year), and he spent the next few years convalescing. He made frequent trips to Italy for warmth, and to the Alps for pure air. To his friends he continued to joke, perhaps too frequently, about his mounting insignificance.[8] Shortly after retirement, however, he began to conduct a series of controversies from the confines of his study. His first antagonists were familiar: Anglican clerics and other Church leaders. The issues – agnosticism, miracles, the Genesis account of creation – were still compelling to some. John Knowles, editor of the periodical *Nineteenth Century*, encouraged the disputes and solicited articles from all sides. Knowles's journal was one of several late-Victorian periodicals

---

[6] See the excerpt from Huxley's Royal Society address, and his letter to Lord Iddesleigh, 24 November 1885, in L. Huxley ed. 1900, 2: 107–9.

[7] T. H. Huxley 1871b: 287.

[8] See, for example, Huxley to Michael Foster, 12 March 1885, in L. Huxley ed. 1900, 2: 98–9.

that aspired, like their mid-century predecessors, to be vital organs of liberalism. The timing of the controversies and the medium in which they unfolded were in some ways felicitous. At a period when Huxley felt his own influence flagging, his status in the scientific community diminished, and his proximity to an idle ruling class uncomfortably close, he could assume the familiar role of social critic, freeing the English people from the oppressive forces of priestcraft, and protecting the language of scientific explanation and natural law from abuse. In a letter to Joseph Hooker in 1889, however, he complained that a fair portion of his scientific colleagues had begun to withdraw from such public disputes:

> I am very glad that you see the importance of doing battle with the clericals. I am astounded at the narrowness of view of many of our colleagues on this point. They shut their eyes to the obstacles which clericalism raises in every direction against scientific ways of thinking, which are even more important than scientific discoveries. I desire that the next generation may be less fettered by the gross and stupid superstitions of orthodoxy than mine has been.[9]

Precisely what was at stake for Huxley in these protracted debates is suggested in his choice of William Gladstone, the target of five articles, as the clerical opponent par excellence.[10] During the mid-1880s, Gladstone, for many years a prolific writer on Christian literature and the Church, took it upon himself to defend the authenticity of miracles. One of the centerpieces of his long debate with Huxley was a passage from the Bible in which Jesus cast out evil spirits from the soul of a Gadarene supplicant into a herd of pigs. According to Huxley, the laws of nature and society were contravened by claims about spirits passing from humans to pigs. A literal reading of this passage would encourage disrespect for private property (the spoliation of a herd of pigs). Thus biblical literalism could prompt irrational beliefs and immoral actions. Huxley had a specific superstitious and criminal population in mind. Gladstone was at this time leading a campaign for Irish Home Rule that Huxley and his circle of scientific friends bitterly opposed. When Huxley referred to the "Grand Old Man" as "the keeper of the herds of swine," he was thus attacking a particular form of leadership and a particular sort of public. But his criticism of Gladstone's biblical hermeneutics was not simply a veil for imperial politics; rather, Home Rule and bibliolatry were both examples of what happened when statesmen abandoned the wisdom of men of scientific learning and curried the favor of "average opinion." While Huxley gathered the evidence of

---

[9] Huxley to Hooker, 22 May 1889, in L. Huxley ed. 1900, 2: 234.
[10] All of the articles appeared in the *Nineteenth Century*: Huxley 1885b, 1886a, 1886b, 1890a, and 1891b.

biblical scholars in what had become the science of textual criticism, and assembled anthropological data on the Irish question together with Lubbock and Tyndall, Gladstone conjured up specious theories that fed the passions and prejudices of the ignorant.[11] In religion and in politics he appealed, in Huxley's words, to "the herd" and, in particular, to its "great heart" rather than its "weak head." "It is to me a grave thing," Huxley wrote to John Skelton, "that the destinies of this country should at present be seriously influenced by a man, who . . . is nothing but a copious shuffler."[12]

Huxley's debates with Gladstone, because they concerned the criteria of good leadership, brought out important assumptions regarding the public that was to be led. Still a prominent M.P. in the late 1880s, Gladstone could not be addressed or dismissed like some of Huxley's other opponents, that is, as the representative of a clerical caste jealous of its elite status. He was among the chief spokesmen of a Liberal Party that was now trying to represent an increasingly discontented body of workers.[13] Gladstone's efforts, according to Huxley, amounted to a copious shuffling because they duped the people by catering to their uninformed feelings and beliefs. His shuffling was grave because, by appealing to popular opinion, he was in fact enslaved to it: "He was born to be a leader of men," Huxley remarked privately, "and he has debased himself to be a follower of the masses."[14] To Albert Grey, a leader of the parliamentary opposition to Home Rule, Huxley wrote a letter that was later printed in the *Times*: "Have we a real statesman? a man of the calibre of Pitt or Burke, to say nothing of Strafford or Pym, who will stand up and tell his countrymen that this disruption of the union is nothing but a cowardly wickedness?"[15] With this letter, Huxley expressed his usual contempt for popular opinion, as well as his enduring suspicion of anyone who would represent it. But this was not the same public ignorance against which he had always defined his cultural role, an ignorance

---

[11] Huxley, Lubbock, and Tyndall maintained that the Irish, like the English, were of a mixed racial stock. The collaboration of men of science on the Irish question is discussed in Di Gregorio 1984: 178–9. See also Lorimer 1988. For an account situating Huxley's later writings in the context of Liberal Unionism, see McGeachie 1990.

[12] Huxley to Skelton, 21 January 1886, in L. Huxley ed. 1900, 2: 122. Skelton was a biographer and an essayist and a frequent contributor to *Fraser's* and *Blackwood's Magazines*; he was involved in Scottish government.

[13] On the efforts of parliamentary Liberals to shape the political form of workers through languages of "class" and "mass," see Briggs 1960 and 1979. On Gladstone's populism, see P. Clarke 1999.

[14] From a conversation "some time in 1887 or 1888," between Huxley and Reverend Benjamin Waugh, with whom Huxley had served on the London School Board, as recorded in Waugh's notes to Dr. J. H. Gladstone, cited in L. Huxley ed. 1900, 1: 353.

[15] Huxley to Grey, 21 March 1886, in L. Huxley ed. 1900, 2: 125.

that was supposed to be educable, a public that was ultimately moved by reason and moral example. Behind Huxley's rhetoric were fresh assumptions about the crowd, that brutish body where self-reliant, free-thinking agents suddenly lost their reason and where England's industrious, commonsensical people were now collecting (and deteriorating) in large, volatile numbers.[16] Likewise, Gladstone was not just a rearguard of outworn tradition or a practitioner of party politics as usual. He was proof to Huxley that democracy was really despotism.

Land, Leadership, and Learning

In the course of his engagements over the interpretation of Scripture, Huxley paused to participate in a *Nineteenth Century* symposium on international copyright law. In an 1887 article, he suggested one way in which Gladstone's politics – giving the people what they wanted – threatened the intellectual estate of men of science:

> It has become an axiom among a large and influential class of our politicians that a want constitutes a good claim for that which you want, but which other people happen to possess. The "earth hunger" of the many has established itself as an excellent spoliation of the landowning few ... The course of action by which ... Transatlantic readers propose to deal with British authors is but another anticipation of that social millenium when the "Have-nots," where they lack land, or house, or money, or capacity, or morals, will have parted among themselves all the belongings of the "Haves."[17]

Such a piece, in which science writers were associated with the landowning few, and transatlantic claims to knowledge likened to millennial dreams and criminal acts, could hardly have been composed by Huxley a decade earlier. His works, particularly his lectures to working men, had been printed and sold without authorization since the early 1860s. His objections to this had been private and short-winded because it was money that seemed to be at issue, a matter to which he was pledged to be publicly indifferent.[18] In an address delivered in 1871 in the wake of the formation of the Paris Commune, Huxley had supported a state tax on private property to fund a national system of education, by posing science as an inexhaustible intellectual commons: "The investigation of Nature is an infinite pasture-ground, where all may graze, and where

---

[16] For a survey of these new scientific discourses on the crowd, see Pick 1989.

[17] T. H. Huxley 1887b: 620–1.

[18] For Huxley's reaction to the printing of his lectures "On Our Knowledge of the Causes of the Phenomena of Organic Nature" by Hardwicke, see his letters to Hooker, 16 January 1862, and to Lyell, 28 January 1862, in L. Huxley ed. 1900, 1: 207–8.

the more bite, the longer the grass grows, the sweeter is its flavour, and the more it nourishes."[19] Such public assertions of the democracy of knowledge were important to the self-presentation of men of science, many of whom had themselves risen from modest positions and whose role was to help others to do the same through learning. By such assertions, men of science claimed a role in more widespread liberal movements: criticism of the Church and paternal government, reforms in the Universities, public and elementary schools, and the construction of an enlightened public sphere through the medium of print. In correspondence during the early 1870s among Huxley, Tyndall, Spencer, and Henry King, publisher of the International Scientific Series, the pirates of scientific knowledge were depicted as crass (a.k.a. American) tradesmen, not communists.[20]

But if Huxley had once represented his work as "organized common sense," if the truths of science were not supposed to rest merely on the authority of its practitioners and if all were at liberty to test them, this was because legitimate speculation was exchanged only between learned men and because learned debate was carried on only before audiences who did not question institutions like private property and entities like inborn genius. Men of science were able to traffic between an order in which social divisions were based on property and one in which such divisions were based on carefully circumscribed abilities, because their abilities were supported in large part by, and were pitched in support of, those with property. In such a society, knowledge remained implicitly the property of an elect. Workers could make themselves intelligent, propertied, and sovereign as individuals of a certain kind; they could not rise as a class. As Chapters 3 and 4 have shown, reforms designed to open science, literature, and education to persons other than polite gentlemen also concentrated authority among new learned elites. As Huxley's reference to "earth hunger" revealed, however, this relationship between private property, intellectual capacity, and the ability to rule was now threatened. The tight circles in which men of learning and means had settled matters of truth and of state were being forced open by those for whom want constituted a just claim.

On 7 November 1889, the *Times* reported on a meeting between John Morley, M.P. for Newcastle, and a group of his constituents to discuss a socialist platform that included among its proposals land

---

[19] Huxley 1871b: 282. This was originally an address to the Midland Institute in Birmingham.

[20] The International Scientific Series ran from 1871 to 1911. Huxley, Tyndall, and Spencer, who formed the original editorial committee for Britain, aimed to secure royalties for men of science (see Howsam 2000).

nationalization. When asked by Morley what was meant by "nationalization," John Laidler, a bricklayer, replied that when the owners died their land should go to town councils who in turn "should deal with it according to the democratic spirit accordingly as they were elected for the purpose." He added "that Mr. Herbert Spencer had said that the land had been taken by force and by fraud." This brief exchange followed. Morley: "Has Mr. Spencer said this?" Laidler: "Yes, we all know." Morley: "You are aware that he has recalled some of the things he has laid down?" Laidler: "If he has stated truth and recalled it the truth will prevail." In the same issue, the *Times* printed a letter from Spencer confirming that he had indeed renounced some of his earlier opinions, particularly those set forth in the work to which Laidler referred, *Social Statics*.[21] Moreover, he had in recent years twice pointed out his renunciations and had stopped any republication or translation of the said work. His original aim, he asserted, had been to refute "socialism and communism" to which he was then "as profoundly averse" as he was now.[22]

The Newcastle debate and Spencer's response were the beginning of a controversy that occupied the columns of the *Times* for the better part of a month. Two days after the original report, the paper printed a letter of remonstrance by the writer Frederick Greenwood to certain "Social Philosophers" and their "ill-considered theorizing" in an "age of popular education" and "social unrest." Thousands, Greenwood cautioned, "are ever on the alert for warranted theories of social reform that will better their condition" and that "habitually hang on the authority of great men."[23] The *Times* directed its own indictment at socialists who pilfered from the works of philosophers just as they would pillage private property: "When a man wants a theory to justify a plunder he is not usually very scrupulous in his treatment of authorities." Spencer reentered a plea that he could not remember everything he had written over forty years, but if he had once lent support to socialism, he now repudiated it. Laidler rejoined: while such retractions were common among politicians, "one does not expect the same vacillation on the part of a distinguished philosopher." Conceding that a redistribution of land would entail complex problems of compensation, he insisted: "there are others besides the landed class to be considered. The rights of the many

---

[21] See especially the conclusion to chapter 9 of *Social Statics*, in which Spencer claimed that "the theory of co-heirship of all men to the soil, is consistent with the highest civilization" (Spencer 1851: 125).

[22] *Times*, 7 November 1889. On Spencer and socialism, see Taylor 1992. See also the collection of contemporary reviews of Spencer's politics in Offer ed. 2000, vol. 4.

[23] *Times*, 9 November 1889. Frederick Greenwood was a novelist, journalist, and the first editor of the *Pall Mall Gazette*.

are in abeyance. If the author would permit it to be reprinted, what an admirable tract the ninth chapter of *Social Statics* would be for the propagation of Socialistic principles."[24]

The *Times* debate was different from any that might have occupied its high-brow columns in the mid-Victorian decades. The bricklayer from Newcastle who addressed Morley was among an increasing number of labor leaders with subversive claims to land, leadership, and learning. Laidler linked the equitable allotment of land to the free circulation of Spencer's text. The workers whom Laidler claimed to represent were creating forms of association – new inclusive unions, a Labour Party, learning societies – that demanded access to wealth and political decision making as well as to the production and distribution of knowledge. Unlike the institutions of artisan learning in the middle decades (e.g., Mechanics' Institutes, the Working Men's College), the study groups in which Laidler might have read and discussed Spencer were organized around a policy of collective assertion.[25] Such communities pursued learning with the understanding that they were improving themselves and laying claim to knowledge that ought to be common but that had been made private to protect the property of a few. For working-class representatives, the debate over land nationalization was in part a struggle between a science that was truly social, that is, applicable and beneficial to society in general, and a science that was produced for selected individuals. Tom Mann, a leader of the Dockers' Union, recalled how he had read the period's most popular text on land reform, Henry George's *Progress and Poverty*, as a work that refuted Malthusian economics by showing that laissez-faire principles did not abide in any place (factories, mines, farms, labor legislation) where authoritarianism protected the privileges of owners.[26]

Mann's reading, like Laidler's, was part of a new culture that had formed around workers' concerns and that was constructed both in opposition to and as the heir of elite scientific culture. Groups such as the Social Democratic Federation (SDF), the Socialist League, and the Fabians were founded in the early 1880s and were all committed in various ways to the development and diffusion of what William Morris called "rational, scientific socialism."[27] It was, in part, through the appropriation of scientific knowledge and scientific ethos by such groups that a bricklayer's arguments were able to enter the forum of the *Times* and

[24] *Times*, 9, 11, 15, and 19 November 1889.
[25] On education for the working classes in the 1880s and 1890s, see Simon 1965 and Laurent 1984. On earlier popular education movements, see Shapin and Barnes 1977, Johnson 1979, and Cooter 1984.
[26] Mann 1896: 79.
[27] Morris 1885: 25. For accounts of these organizations, see Hobsbawm 1964.

not be summarily dismissed. Socialism, wrote H. L. Hyndman, founder of the SDF, was now a "distinct scientific historical theory, based upon political economy and the evolution of society." In his popularizations of Marx, Hyndman turned the agenda of liberal social science on its head. Struggle was not between individuals, but between classes. Progress was toward popular sovereignty. Nature would be mastered to benefit all equitably.[28] Bernard Shaw and Sydney Webb among the Fabians used the virtues claimed by men of science – disinterestedness, public spirit, creativity rather than acquisitiveness – to fashion their own "intellectual," "professional," and "literary proletariat."[29] Central to the formation of this counterculture were separate spaces of sociability and material production. Socialists were inclined to publish not in Morley's *Fortnightly Review* or other standard-bearers of liberalism, but in their own papers like *Justice* and the *Commonweal*. They did not frequent the Athenaeum drawing rooms. They could not attend meetings of the famous Metaphysical Society, for it had dissolved in 1881, and not perhaps, as Huxley once suggested, from "too much love."[30]

If such sharp divisions between classes and within the circles of learning had not existed since the demise of Chartism, this was not only because the middle decades were prosperous but also because the sciences, which were presented as models of rational activity and as sources of political wisdom, had been organized to prevent the problems of sovereignty from being posed in a radical way. In response to the bricklayer, Morley, Spencer, Greenwood, and the *Times* editor all suggested that if the ownership and management of land by the people was disastrous, so was the control of knowledge. To some degree, their judgment derived from a long-standing political tradition in which franchise reforms were resisted as concessions to the impulsive and unreflecting classes. In their favor, however, Laidler's opponents possessed more than an anecdotal knowledge about assemblies of masses, their irresistible force and violent, (un)predictable behavior. They could draw on highly scientized (medical, sociological, anthropological) discourses about races, criminals, cities, and civilization that carefully classified the bricklayer and his demands. By employing the concept of degeneration with increasing precision, men of science could explain workers' collectives and blind, indeterminate behavior as savage reversions, nervous diseases, or the irrational movements of bodily automata.[31] Such

[28] Hyndman 1884: 4.

[29] Shaw and Webb, cited in Hobsbawm 1964: 258.

[30] Huxley's remarks on the Metaphysical Society are recounted in L. Huxley ed. 1900, 1: 316.

[31] For contemporary accounts of degeneration, see, for example, [Anonymous] 1886, 1889b, and 1889c.

explanations prevented voices like Laidler's from being heard in concert with that of an M.P. or a philosopher.

Despite their proliferation in the 1880s, these discourses could not forestall political debate. This was in part because workers did not have to rely on them for self-understanding and self-help. The locus of authority, both in the body politic and in the social sciences that helped to constitute it, was now an open question. Thus neither parliamentary Liberals, nor social theorists, nor the *Times* editor who favored them could decide the issue of land nationalization among themselves. Yet despite its condescending tone, the *Times* seemed to revel in displaying the workers' perspective to the disadvantage of a man of learning. Laidler's remarks called into question the authority of "great men" of science. Spencer's status as a philosopher may have won him notice among workers, but his efforts to control the political significance of his early work by suppressing its republication were unsuccessful. Soon the English Land Restoration League would distribute *Social Statics* in leaflets to the rural poor.[32] If science could be used to negotiate the rights of workers, who would represent science?

On 12 November, Huxley entered the fray, condemning Spencer's "absolute political ethics" and "a priori politics" and offering in their stead his own "common morality and common sense." While Spencer pleaded forgetfulness, the *Times* editor waxed smug, and Morley drew applause from an audience of businessmen by proclaiming socialism "against human nature," Huxley voiced concern that Laidler's arguments had not been adequately addressed:

> I conceive it to be a matter of vital importance to the whole nation that the representatives of labour should be under no misapprehension with respect to the grounds of any action they may think fit to take. And as all my life I have done my best to bring sound knowledge within reach of the working classes I trust that they will do me the justice to believe that I am actuated by no other motive now.[33]

By reasserting the role he had played in the middle decades as the instructor of working men, Huxley tried to prevent a Newcastle worker from appropriating the intellectual property of a friend and from degrading that friend, a philosopher, by making him a "representative," rather than an educator, of labor. Spencer's social science and his dignity as a "great man" above party politics and class interest had been used by the bricklayer to undermine the society that philosophers and

---

[32] These activities of the English Land Restoration League are described in a letter from Thomas F. Walker to the editor of the Birmingham *Daily Gazette*, 27 August 1894.

[33] *Times*, 12, 18, and 21 November 1889. See also the report of Morley's speech at the "80 Club," *Times*, 20 November 1889.

men of science had worked to build. Spencer's ideas had been separated from his alleged interests, and Spencer himself had been ranked among the party politicians. Laidler's action was humiliating, because it was exactly what men of science, in their own self-fashioning and rise to eminence, had done to women and to other classes and learned groups. His criticism was of a kind that was not supposed to come from outside the circles of learning or to fall upon those with specially endowed powers of mind.

Against these recriminations, Huxley had to defend the economy of self-interest, his own disinterestedness as a man of science, and the integrity of a scientific community that seemed to be dividing and disputing over knowledge like so many political factions. Situated between more extreme proponents of eugenics such as Karl Pearson and Francis Galton, and more socialistic meliorists such as Alfred Russel Wallace (co-inventor of evolution by natural selection and the recent founder and president of the Land Nationalisation Society), Huxley could not avoid crossing swords with Spencer, even to the point of jeopardizing their friendship.[34] Despite repudiating his own earlier work on socialism, Spencer had continued to furnish ammunition for his opponents. In the *Times* debate, he had quoted from his recent *Principles of Sociology* to the effect that private land was a prisoner of war seized by "dominant men ... during the evolution of the militant type." "It seems possible," the passage concluded, "that primitive ownership of land by the community ... will be revived as industrialism develops."[35]

As the *Times* debate came to a close, Huxley began a series of four articles that appeared in the first half of 1890, attacking socialism and the (singular) working class as aberrations of nature. Here, in the more exclusive medium of Knowles's *Nineteenth Century*, he asserted that men were naturally unequal and that inequalities of intellect, being the most fundamental of all, were the basis of those in economics and politics. He answered the land nationalists with arguments that private property was the product not of force but of labor, that capital was not the enemy but the "mother of labour," and that intellect was the highest form of capital, begetting all others: "The witless man will be poverty-stricken in ideas, the clever man will be a capitalist in that same commodity, which in the long run buys all other commodities; one will miss opportunities, the other will make them; and, proclaim human equality as loudly as you like, Witless will serve his brother."[36]

[34] On Galton and Pearson, see Kevles 1985. On eugenic policies in Britain, see Searle 1976. Wallace's socialism is discussed in Durant 1979.

[35] *Times*, 15 November 1889.

[36] Huxley 1890d and 1890b: 109. See also Huxley 1890c and 1890e.

Huxley had originally planned to publish these pieces in a pamphlet after the fashion of his working men's lectures on natural science: "a sort of 'Primer of Politics' for the masses."[37] That he chose instead a journal that was not popular among the audience whom men of science claimed to educate was indicative of his distaste for a public sphere that was no longer middle class or polite. High-brow forums like *Nineteenth Century* had become safe havens where learned men could display their concerns for workers' movements without actually engaging with workers. Huxley, together with many men of science and liberal politicians, refused to concede that a singular "working class" actually existed, and this, as much as their own authority, was what was at stake in these debates. In uneasy conjunction with their discourses on atavism and decay were others that continued to address workers as if they were the "working men" of old. Organized under the rubric of "technical education," such disciplines sought, through an appeal to reason and improvement, to elevate those workers "of a better class" above the rest, as the franchise reforms had done.

Arming for War

On 29 November 1887, Huxley made an eight-hour journey to Manchester to deliver a speech before the National Association for the Promotion of Technical Education. He considered the event of great importance, for not only were public appearances difficult for him but on this occasion he was forced to leave his wife just ten days after the death of their daughter, Marian. He wrote to Hooker shortly before his departure, "I cannot leave them in the lurch after stirring up the business in the way I have done, and I must go and give my address. But I must get back to my poor wife as fast as I can, and I cannot face any more publicity than that which it would be cowardly to shirk just now."[38] The association that Huxley addressed had been established by Henry Roscoe, the professor of chemistry at Owen's College. Its aims were to systematize and institute, on a national scale, schemes of practical instruction that had developed locally and haphazardly for several decades. Huxley had been involved with other men of science in the planning and promotion of these schemes since he first began teaching at the School of Mines in the mid-1850s.[39] The themes of Huxley's 1887

---

[37] Huxley to Hooker, 14 December 1889, in L. Huxley ed. 1900, 2: 245.

[38] Huxley to Hooker, 21 November 1887, in L. Huxley ed. 1900, 2: 180.

[39] On the technical education movement in Britain, see Roderick and Stephens 1972 and MacLeod 1977.

speech were quite different from those he had emphasized during the several decades after mid-century, however.

The technical education that Huxley and others like Roscoe, Tyndall, and Playfair had promoted in the past was an effort in part to train British workers in specific occupations and in part to establish among a practically oriented British public the practical value of science. This effort was widely supported by men of science eager to take credit for the progress of industry. They were encouraged by the British Association for the Advancement of Science and by *Nature* magazine in order to create space for scientific professions.[40] But technical education meant much more to men of science than just career building. Technical education was designed to show how science was woven into everday life, how it was used, or ought to be used, by every person to solve basic probems and perform basic operations. Finally, it was intended to preserve the elevated status of "pure" or "theoretical" pursuits that, though allegedly responsible for every useful innovation, were driven by higher motives than application. It was to achieve this last goal by being preceded by, and founded upon, a liberal education. For men of science, technical education was thus a movement to integrate the lives and activities of all workers, intellectual and manual, to induce progress and ensure social stability and, at the same time, protect the status of science as elite culture. It was designed to be a program by which men of science were brought into intimate relations with industry, as well as a means by which they could exercise authority over industry.

There were tensions in the movement from the very beginning. Men of science did not easily convince statesmen and industrialists of the value of liberal education for workers, or of "pure" science for technical practice. Some in politics continued to view any form of education as inappropriate (or dangerous) for persons whose lot in life was hard labor.[41] Workers themselves, for the most part, did not seem to appreciate either the higher values of elite culture or the practical benefits of theoretical science. If they sought technical education, it was to raise their pay. If they sought liberal education, which was rare (and rarely available to them), it was to raise their station. These differences about the aims of technical education brought to the surface discrepancies among different social groups over the nature and value of science. In defending their own status as members of a high cultural order, men of science sometimes implicitly criticized the economic system that their technical education programs were supposed to serve. When asked by

---

[40] For programatic statements by men of science on technical education, see *Parliamentary Papers, 1867–8*, vol. 54, Playfair 1870, and Lockyer 1906.

[41] See, for example, Armstrong 1888a and 1888b. See also Lockyer 1888 and Playfair 1888.

a commissioner on the Select Committee on Scientific Instruction in 1868 whether his proposed course of social science for workers would "obviate the ignorance which tends to strikes," Huxley indicated that *all* classes needed to comprehend the conditions of human welfare: "I should like to be understood to mean strikes on both sides; strikes of the masters as well as strikes of the men; I wish to be quite impartial upon that point."[42] Precisely what Huxley meant by strikes on the part of masters, whether lockouts, wage cuts, layoffs, or even blows, is not clear. It is certain, however, that the science Huxley deemed integral to technical education was not there to encourage drastic social reforms. It was designed to function, though by other means, in the same way as some of the conscience-raising sermons of clerics: to temper the harshness of a system of private ownership and individual competition in which those who had wealth were strongly favored. Its aim was to instruct owners on the responsibilities of wealth as well as to teach operators the duties of labor. The differences between men of science, industrialists, and statesmen were thus largely matters of emphasis. These differences did not prevent the building of a variety of institutions in the mid-Victorian period oriented toward practical learning. Among these was Huxley's own School of Mines, which had been established together with the Royal College of Chemistry and the Geological Survey under the Board of Trade. One of the chief functions of the school was the training of teachers for courses in applied sciences, most of which were sponsored locally and privately. Huxley prepared many of these teachers. He was also an examiner for the new Science and Art Department, which loosely administered these courses through a system of examinations, and which paid the teachers according to results. Shortly after Huxley's tenure began in 1854, the school commenced its own series of "Lectures to Working Men."

What these institutions shared, along with many others that were constructed during the same period in London and in the provincial towns, was a commitment to training an artisan public to be more productive, a desire to bring different classes together in the common enterprise of improvement, and a general tone of optimism. As late as 1877, Huxley could still address such a mixed audience at the Working Men's Club and Institute Union in London, posing science as a kind of "handicraft" and likening his own specialty, comparative anatomy, to manual labor. He could offer technical education as an instrument of self-help, enabling the worker to fare well in practical life. He could also counsel masters not to be jealous of elevating their workmen, for technical instruction would not induce workers to idleness or discontent. It would fit them

[42] *Parliamentary Papers, 1867–8*, vol. 15, p. 402.

for industry and allow them to realize what was their respectable ambition: "to go through life with moderate exertion and a fair share of ease."[43]

In theory, the only difference between these technical education efforts and those of the 1880s was scale. By 1883 the City and Guilds of London Institute, established ten years earlier by various metropolitan companies, had grown from 184 students to over 5,000. It offered certificates in cotton production, gas and steel manufacturing, carriage building, and agriculture. With the School of Mines, it sponsored a teacher-training facility in South Kensington.[44] Similar schools and colleges, financed by local industries, were in every industrial town. In honor of the Queen's Jubilee, the government made plans with London companies for an Imperial Institute, to include an exhibition and lecture hall, a gentlemen's club, and a meeting space for improving workers and capitalists, with a special orientation toward colonial enterprise. The National Association for the Promotion of Technical Education, which Huxley addressed in Manchester, was another joint venture of businessmen, politicians, and men of science. It sought to organize all local efforts by placing them under the school boards, supervised in turn by the Science and Art Department, and drawing funds from municipal rates.

Huxley's speech to the National Association indicated, however, that this grandness of scale and coordination was accompanied by a significant change in tone. In his "Technical Education" address of 1887, and in other remarks on the subject from the same period, Huxley reiterated his old agenda: broad elementary education to be followed by special training in Science and Art Department courses and factory schools; a scholarship program for the exceptionally gifted; and social science. He supported the initiatives of private industry and encouraged their further organization and financing under the direction of the state and local governments. But missing from these utterances was the optimism that had been so characteristic of the earlier movement, and with it the expectation that workers who followed this system of education and applied themselves thereafter were bound to better their condition. In place of the old rhetoric of self-help and improvement was a new rhetoric of continuous and inevitable war. That the warfare was economic scarcely mitigated its harshness:

> The terrible battle of competition between the different nations of the world is no transitory phenomenon ... It is the inevitable result of that which takes place throughout nature and affects man's part of nature as

[43] Huxley 1877b: 405–6, 422.
[44] On the City and Guilds of London Institute, see J. Lang 1978 and Stevens ed. 1993.

much as any other – namely, the struggle for existence, arising out of the constant tendency of all creatures in the animated world to multiply indefinitely. It is that . . . which is at the bottom of all the great movements of history . . . that inherent tendency of the social organism to generate the causes of its own destruction . . . We are at present in the swim of one of those vast movements in which, with a population far in excess of that which we can feed, we are saved from a catastrophe, through the impossibility of feeding them, solely by our possession of a fair share of the markets of the world.[45]

While Huxley was composing this and other expositions of technical education in the late 1880s, he was also writing an obituary notice on Darwin for the *Proceedings* of the Royal Society. Though he undertook this piece in 1883, he did not complete it until five years later.[46] In letters to Foster and Hooker in early 1888, Huxley remarked that he was still rereading the *Origin of Species*, trying to separate the "substance" of the theory from its "accidents," with the aim of warding off a generation of "hostile comments and would-be improvements." Even though he had written at least a half-dozen abstracts of the work and was reading it, he said, "for the nth time," he was "getting along slowly" and finding it "one of the most difficult books to exhaust that ever was written."[47] At this juncture in his life, it seems that Huxley had difficulty concluding what he had always concluded previously about Darwin's theory: that its points of central importance were the facts of variation, the Malthusian principle of overpopulation, and its consequence, universal struggle. As Huxley finally came round to saying once again in the obituary article, it was immaterial how organisms differed from each other or why. For Darwin's theory to work, it only mattered that organisms turned these differences against each other in a struggle to survive.[48]

This conclusion, however difficult to reach, had its advantages: for if the technical education movement in which Huxley was currently engaged could not bring the classes of England together through the common business of improvement, it could at least unite them through the common labor of war. Economic war was simply the form that the natural struggle for existence took in human society, a struggle that arose out of the tendency of humans to reproduce beyond the limits of their food supply. Huxley could now claim that the issues of technical education – economic production, class relations, and social welfare – were

---

[45] Huxley 1887a: 446–7.

[46] Huxley 1888a.

[47] Huxley to Foster, 14 February 1888, and Huxley to Hooker, 9 and 23 March 1888, in L. Huxley ed. 1900, 2 : 190–3.

[48] Huxley 1888a: 287–8.

"purely scientific ones." In his 1888 article "The Struggle for Existence in Human Society," written for the *Nineteenth Century* symposium on technical education, he applied the equation of the *Origin* systematically. The great movements of history and the great crises of the day could all be addressed through the application of simple biological truths: the tendency of living things to reproduce without limit, the restricted dimensions of habitable land, and the ensuing contest for existence.[49] This struggle (it could now be told) was not merely or even primarily between individuals, but between groups. Huxley thus showed how evolutionary theory could help liberals to modify their individualism and come to scientific terms with the problems of mass society. He could mobilize the dissident English masses for international competition and sell science as the latest and best form of arms.

This solution came at a considerable price. Perhaps Huxley's most bellicose call to industrial war came in a speech at Mansion House to raise subscriptions for the Imperial Institute, the proposed center for the advancement of commerce in the colonies. The *Pall Mall Gazette* reported his words like this: "It was a hard thing to say, but . . . industrial competition among the peoples of the world at the present time was warfare which must be carried on by the means of modern warfare, namely scientific knowledge, organization, and discipline."[50] When contributions began to flag due to disagreements over the site, Huxley received letters from a member of the London Chamber of Commerce, Thomas Christy, urging him to represent the interests of "city men": "The Merchants and Bankers will be guided by your advice."[51] Huxley proceeded to write several letters to the *Times*, each more strident than the last: "our prosperity is threatened as it has never been threatened before. Germany and America, full of material resources, of scientific knowledge, and of industrial skill, have learnt all we have had to teach them, and have barred the way of our trade by restrictive tariffs. They are industrially at war with us; are we to arm to meet them, or not?" Technical education, he advised, was a "war tax for the purposes of defence." The Imperial Institute was "an intelligence department for the army of industry."[52] Huxley did not have to support the mobilization of troops and munitions to be deeply implicated in the increasingly violent practices of imperialism. His own discourse declared that there was no difference between war and commerce. Huxley would continue to urge that employers not sacrifice the "dignity" of their workers to the altar of

[49] Huxley 1888b.
[50] *Pall Mall Gazette*, 15 January 1887.
[51] Christy to Huxley, 20 January 1887, 12 and 17 February 1887, HP: 12.189–91.
[52] *Times*, 20 January 1887, 17 and 22 February 1887.

competitiveness; that they not establish wage rates and work conditions that "degraded" their employees.[53] But according to his group-selection model of human evolution, terms like "dignity" and "degradation" had little significance except as part of a bioeconomic equation of fitness. Social welfare, self-sacrifice, civic spirit – the virtues that men of science had always claimed to stand for – now existed principally as a means of waging more effective international competition. The only rationale Huxley could evoke to shame the bad employer was that a degraded worker was a discontented worker, and a discontented worker was an unproductive worker (or worse). At the same time, his military rendition of commerce sanctioned severe burdens on workers, who in a world at war were expected to make greater and greater sacrifices.

If Huxley's developmental theories could not negotiate consent among elites, or command the public sphere on issues of religious belief, home rule, land nationalization, and copyright law, neither could they succeed in the field of technical education. Leveling models of evolution and critiques of ineluctable struggle, popular among radical circles during the first half-century, emerged again.[54] The same *Nineteenth Century* symposium that carried Huxley's article "The Struggle for Existence" published a piece by the Russian anarchist Peter Kropotkin, arguing for a form of technical education that would reverse the divisions of labor that separated "the manual worker and the intellectual worker... to the detriment of both," by teaching science and handicrafts to everyone from an early age and rendering each person "a producer of both manual and intellectual work."[55] In response to Huxley's Imperial Institute address, the *Pall Mall Gazette* contributed a more sarcastic tone to the discussion: "One would humbly ask Professor Huxley if science and industry are for the good of humanity, or humanity for the good of science and industry."[56] The alliance between science and industry that had been asserted during the middle decades by scientific elites, often in the face of opposition from men of business, was now taken for granted, only to have its moral integrity assailed. This alliance had to be defended against a variety of critics, and in media that did not discriminate among contributors as the prevailing organs of mid-Victorian liberalism had done. The editor of the *Pall Mall Gazette* during the 1880s, William Stead, used his columns to follow Salvation Army officers on

---

[53] Huxley, evidence before the Royal Commission on Technical Education, 1882 (see *Parliamentary Papers, 1884*, vol. 31, pp. 322–4). The links between evolutionary theory, militarism, and war are explored in Crook 1994 and Brantlinger 1997.

[54] On radical uses of evolutionary theory in the first half of the nineteenth century, see Desmond 1987 and 1989.

[55] Kropotkin 1888a: 497–9. See also Kropotkin 1888b and 1888c.

[56] *Pall Mall Gazette*, 15 January 1887.

lurid incursions into London brothels.[57] Soon he would ghostwrite a book for William Booth that displayed a lively irreverence for scientific elites and the ground they had staked out in previous high culture wars, and that mingled social science, religious enthusiasm, and public instruction in a military campaign to rival Huxley's own.

The General's Scheme

In February 1888 the Salvation Army opened a sleeping shelter near the West India Docks, offering soup, bread, music, and a meeting for a penny. It was the first of many such establishments in the poor districts of London, and only part of a growing relief network that included "rescue homes" for women and "prison-gate brigades" for released convicts. By 1890 the Army had organized all such programs under its new Social Reform Wing and, having reduced prices by one-half during the dockers' strike, was able to report that some 250,000 shelter tickets had been sold in the previous year.[58] Together with his wife Catherine, William Booth had founded the Army in the mid-1860s, shortly after the Methodist New Connection, with whom he had served for ten years, asked him to give up revivalism for a stable congregation. Army meetings were bodily dramas of enlightenment and surrender, with hearts melting to boisterous music. A peculiar combination of patriarchy and egalitarianism, militarism and freedom of expression, staid middling morality and the boom and bustle of workers' culture, the Army had a complex reception among different groups of Victorian society. Converts, largely from the urban working class, found opportunities to participate and preach that were rare (especially for women) among Christian denominations, though the Army's campaigns against profligacy, insobriety, and idleness aroused considerable working-class ridicule.[59] In her book *The Salvation Army in Relation to the Church and State*, Catherine Booth described how the work of the organization would help "to preserve the country from mob-violence and revolution."[60] If the Army's fervor and sordid exposés were occasionally embarrassing to the respectable, its attempts to make the masses more middle class were generally welcomed by a liberal state increasingly taxed by the swelling numbers of poor in the cities. Its press campaigns

[57] Sandall 1955, 3: 30–40.

[58] Sandall 1955, 3: 68–74.

[59] For accounts emphasizing the class and gender complexities of Salvationism, see Briggs 1981 and Walker 1991: 92–6.

[60] C. Booth 1883: 1–4.

in the *Pall Mall Gazette* against prostitution and white slavery lent extensive support to the Criminal Law Amendment Act, which, among its other measures, opened private dwellings to government inspection without a warrant if the occupants were suspected of harboring young women.

As the Army grew, however, so did the ranks of its enemies. The early 1880s, its years of greatest expansion, were marked by riots in some sixty towns, chiefly sparked by brewers and publicans hurt by temperance campaigns. In these conflicts the Army frequently had the legal and economic systems turned against it. Owners warned their employees that those who converted would be out of a job. Mobs commissioned by local tradesmen were given license by police, and civic authorities passed ordinances banning Army processions and open-air meetings.[61] At the same time, years of revival work in the poorest parts of the country began to suggest to Booth that most of the women and men whom he visited or who came to him without property and work could not be reached, and that the material conditions that demoralized them were unjust. Booth's secular optimism, his insistence upon extending the liberal hopes of improvement to all people, led him to criticize capitalism sharply, especially its resigned and cynical, but nonetheless scientific and self-righteous, culture in which poverty and unemployment were inevitable, endemic, and deserved.[62]

By the mid-1880s the Army found itself side by side with socialists in claiming the virtues of social respectability for the whole "class" of workers and in defying the power of political and intellectual authorities to define and enforce certain fundamental Victorian values. Workers' associations and popular resistance movements had enduring ties with religion in England. The first Labour Party representative, Keir Hardie, promised to reclaim Christ for the people and to restore a world in union under God. Tom Mann preached at a Labour church whose minister was a Fabian.[63] Just as the new unionism of the 1880s and 1890s drew extensively on Christian traditions for its ideals of emancipation and brotherhood, and for its language of transforming enthusiasm, Booth, in turn, drew on the resources of the labor movement, its anger, arguments, facts and figures, and practical solutions, for his own program of social reform. Alongside his usual conversion subjects, the

---

[61] Bailey 1977.

[62] Sandall 1955, 3: 65–71.

[63] On Keir Hardie, see McLeod 1984: 83. On Tom Mann, see Hobsbawm 1976: 141–4, 190–2. In the Marxist historiography that long dominated British labor history, religion was often regarded as a "primitive" form of social organization and resistance. See, for example, E. P. Thompson 1963: 31–45, 363–88. For a revisionist approach, see E. Yeo 1987.

criminal, intemperate, and idle crowds of poor, were less conservative targets of reform, the frugal, self-reliant, and commercial middle class. Accompanying his persistent emphasis on the moral state and religious belief of the individual was an elaborate platform for the regeneration of society (see Plate 9).

When *Darkest England* was released in October 1890, all ten thousand copies were sold the first day, and forty thousand more in the first month.[64] Sales may have been encouraged by the title, which Booth lifted from the explorer Henry Morton Stanley's popular work on Africa, published the previous summer. Booth opened the book with a tour through London's hellish depths and dank, unlit streets and depicted their "dwarfed and de-humanized" inhabitants. His descriptions owed much to the print that had flowed fairly continuously since the 1830s, when charity workers, statistical societies, and investigative journalists began their long journeys through the urban underworld.[65] Their narratives of the metropolitan jungle, mingling with natural historical writing on race and species, had made passage from the animal to the savage to the laboring masses and back a commonplace. Favored by liberal moralists and politicians, such discourses tended to project urban problems away from Britain's civilized essence onto degenerate individuals or carefully circumscribed groups. Booth's formulation of society's ills, and his proposed cures, which included mass emigration, factory discipline, and the production of home-loving, hard-working souls, were much indebted to liberal values and institutions. But in *Darkest England*, Booth also placed the responsibility for earthly damnation in dialectical relationship to material conditions. He evoked Stanley's account of slave traders capturing whole villages for forced labor, to depict the behavior of Britain's owners toward the nation's working population:

> Those firms which reduce sweating to a fine art, who systematically and deliberately defraud the workman of his pay, who grind the faces of the poor, and who rob the widow and the orphan, and who for a great pretence make great professions of public spirit and philanthropy, these men nowadays are sent to Parliament to make laws for the people... Read the House of Lords' Report on the Sweating System, and ask if any African slave system ... reveals more misery.[66]

More threatening, however, than this litany of private and public grief was Booth's promise to deliver the poor from "the inferno of their present life." Unreservedly apocalyptic, he pledged to lay down "a true and practical application" of the words of the Hebrew prophet: "Loose

[64] Sandall 1955, 3: 85.
[65] For a survey of this literature, see E. Yeo 1971.
[66] W. Booth 1890: 14.

Plate 9. The Salvation Army scheme. The victims of vice and poverty, saved from the sea by Army officers, begin their ascent through establishments for industrial labor (from W. Booth, *In Darkest England and the Way Out*, 1890).

the bands of wickedness; undo the heavy burdens; let the oppressed go free; break every yoke...bring the poor that are cast out, to thy house...Then they that be of thee shall build the old waste places and Thou shalt raise up the foundations of many generations."[67] Because of the prominence of radical Christian discourses in shaping England's revolutions, the social meaning of this language was unmistakable. Such language was appropriate for Booth because he claimed that the bonds that salvation would loose lay, in part, in the present social system. As the practical application of such leveling prophecies, Booth's welfare organizations undermined the liberal establishments they copied. Confessing deep sympathy for socialists such as Henry George, Booth proceeded to break his own vow to remain silent on the extension of his scheme to the more fortunate: "I don't see how any pacific readjustment of the social and economic relations between classes in this country can be effected except by the gradual substitution of co-operative associations for the present wages system."[68]

The kinship of Booth's scheme with socialism and the labor movement was further sealed through the defense he was compelled to make against a scientific culture in which the salvation of the poor and the cure of social problems through cooperative work were a violation of natural laws of economy and development. Here Booth's radical Christian repertoire enabled him both to invert the familiar critique leveled by men of science against their clerical opponents and to construe prevailing economic and evolutionary theories as the reworking of the oppressive reprobation dogma of the Church. But Booth's criticism extended beyond certain scientific theories to established models of intellectual practice and community in which the scientific study of society and human nature could not be performed by one such as he, the leader of a vulgar religious body whose loud music, rousing meetings and processions, and convulsive conversion experiences forestalled all reflection. In the opening pages of his book, Booth tried to establish that someone from outside the circles of learning could make sound claims to knowledge and that the women and men whom he led with enthusiasm were as rational as the middle-class public that attended British Association meetings or read Knowles's *Nineteenth Century*.

> In this and subsequent chapters I hope to convince those who read them that there is no overstraining in the representation of the facts, and nothing Utopian in the presentation of remedies. I appeal neither to hysterical emotionalists nor headlong enthusiasts; but having tried to approach the examination of this question in a spirit of scientific investigation, I put forth

[67] W. Booth 1890: 19, 4.
[68] W. Booth 1890: 101, 111, 115, 270–1.

my proposals with the view of securing the support and co-operation of
the sober, serious, practical men and women who constitute the saving
strength and moral backbone of the country.[69]

This scientific language was not used figuratively. The possession of
scientific knowledge and methods was as important to Booth's scheme
as enthusiasm. Indeed the two were conjoined by him in ways that both
imitated and undermined the ethos of scientific practitioners. Just as
there was no place in liberal models of society for the "working class"
as a whole to rise, so there was little place in elite codes of scientific
practice for those without specially endowed powers of mind to sustain
well-directed passions for the pursuit of truth.

Booth's "investigations" were also subversive because of where they
were conducted and who implemented them. While drawing on Charles
Booth's *Life and Labour* and other contemporary studies of the urban
poor, the general relied much more heavily on reports gathered by his
own Army officers, sometimes in conjunction with labor leaders, and
on the personal accounts of thousands of women and men whom he
or his officers met through relief work. The vulgar and ignorant were
not only enlisted in the Army ranks and in support of its activities, they
were asked to engage in scientific inquiries. The book became to some
degree a forum for the views of the poor on the nature and causes of
their despair, often in their own words. He prepared his readers for this
passionate induction with an attack on scientific elitism:

> But after all, more minute, patient, intelligent observation has been de-
> voted to the study of Earthworms than to the evolution, or rather the
> degradation, of the Sunken Section of our people. Here and there in the
> immense field individual workers make notes, and occasionally emit a
> wail of despair, but where is there any attempt even so much as to take a
> first preliminary step of counting those who have gone under?[70]

In *Darkest England*, Booth described how his cooperative scheme
would begin to generate its own knowledge. Alongside the economic
and legal institutions that were to be installed in his labor colonies
were "intelligence departments" to produce and distribute an "index
of sociological experiments": "a kind of University in which the
accumulated experience of the human race will be massed, digested,
and rendered available to the humblest toiler in the great work of social
reform."[71] At a meeting in Exeter Hall a month after the publication
of his book, Booth described how scientific spirit was combined with
the spirit of enthusiasm and reform. His scheme, he said, was based

[69] W. Booth 1890: 17.
[70] W. Booth 1890: 20–1.
[71] W. Booth 1890: 282–3.

on "incontrovertible facts"; for his information on incomes, expenses, and operating conditions had been gathered during the previous year by his officers in conjunction with Tom Mann of the Dockers' Union. The success of the dockers' strike that year was due partly to Army relief work. Booth's own politics, he said in the same address, were the "good of the people." His Salvation Army might fairly be described as a new "Socialistic Party."[72]

"A Fair and Adequate Trial"

After his book was published in October, Booth embarked on a speaking tour, making special pleas to City Companies and other corporate bodies for funds and requesting of churches that they take contributions, and of newspapers that they open their columns for subscriptions.[73] The *Times* tracked donations and provided a forum for readers' assessments. Throughout November, favorable letters came in from a variety of sources. The Prince of Wales; the Queen; the earl of Derby; the marquis of Queensbury; the chairman of the London, Brighton and South Coast Railway; and William Gladstone – to most of whom Booth had sent copies of the book – all pledged either their moral or their material support. Less stately advocates included the Dockers' Union, which thanked the general for proposing a practical scheme and offered its services: "Any assistance that we are likely to render will be gladly given, our opinion being that men being made useful citizens is the highest duty of the well-wishers of the nation."[74] Clergymen also supported the plan in large numbers. Cardinal Manning concurred with the general's criticisms of political economy and the poor laws, and he expressed his entire sympathy for the program: "how completely my heart is in your book." The bishop of Manchester pledged one hundred pounds and reportedly envied Booth for employing "religious means which I confess I could not use myself." Frederick Farrar, archdeacon of Westminster and part of the circle with whom Huxley and other men of science had worked closely on educational reform, delivered a sermon paraphrasing Booth's arguments, citing his figures, and chastising the ineffectual churches for being "too nice and cultured."[75] On 10 November, the *Times* gave notice of a circular that had been widely distributed to British clergy

---

[72] Booth's speech, as reported in the *Times*, 18 November 1890. On the Army relief work undertaken in conjunction with the dock strike, see Sandall 1955, 3: 68–70.
[73] *Times*, 5 November 1890.
[74] *Times*, 12 November 1890.
[75] *Times*, 13 November 1890.

calling for the "fair and adequate trial" of Booth's proposals and signed by representatives from a broad range of denominations.[76]

The multiple meanings of salvation in Booth's book facilitated this wide appeal. Invoking a Christian tradition of charity, and integrating spiritual conversion with earthly improvement, Booth offered a patently religious solution to social problems during a period when even the more evangelical denominations were failing to attract urban workers. As their letters indicated, clerics could share in the work of a popular leader and reformer by supporting Booth, even if their own codes and congregations could not abide his approach. Encouraging words from the wealthy suggested that Booth's scheme was a balm to concerns about England's "dangerous classes," and the voluntary nature of his organization satisfied their preference for old-fashioned philanthropy as opposed to more repressive or state-financed measures. But by foregrounding the discontent of workers and the injustices of the wage system, and by promising a form of respectability to a whole outcast and subordinate population, Booth also gained the endorsement of labor leaders and socialists. Successfully cultivating audiences of urban workers and the poor through an auspicious blend of science, social work, and religious enthusiasm, Booth presented at least a potential challenge to certain ruling elites and their established forms of authority. Notably absent from the initial supporting cast were any elites of science. When Huxley began his letters to the *Times* several months after the book's release, the real subversiveness of the general's scheme for this particular audience became clear.

Huxley's entry into the debate was occassioned by a letter from a woman appealing to him for advice on her proposed one-thousand-pound donation. Prompted by this wealthy philanthropist, Huxley became an immediate authority on Booth's plan and Army. His criticism was twofold. By weaving together accounts of the religious emotions with more recent discourses on degeneration, he was able to attack both the principles of individual and social rejuvenation that informed Booth's work, and the method and style of Booth's leadership. Huxley found support in recent works of anthropology and sociology of religion by James Fraser, Friedrich Max Müller, and Andrew Lang, which denigrated religious enthusiasm and its attendant Utopianism as primitive while praising the higher, calmer emotions like wonder and reverence.[77] He added to these social-scientific conclusions the lessons

---

[76] These denominations included the Church of England, the Methodist New Connexion, the Congregationalist Union, the Presbyterian Church, the Baptist Union, the Society of Friends, and the Wesleyan Conference.

[77] See, for example, Max Müller 1873: 103–5. For an account of this literature, see Wheeler-Barclay 1987.

of English history, evoking some of the same affiliations between the Army and radical Christian groups – Puritans, Ranters, Levellers – that Booth had exploited to different ends. Bolstering these critiques with biology, Huxley attributed the savagery of modern cities to marauding bands that had forsaken their Malthusian duties: "hordes of vice and pauperism," he told the *Times* readers, "will destroy modern civilization as effectually as uncivilised tribes of another kind destroyed the great social organization which preceded ours." By contrast, "[an] honest, industrious, self-restraining" population would make even the worst social state prosper. By giving free reign to the passions and, even worse, by fanning the millenarian excitements of religion, Booth had created "the raw materials for a Socialistic Army," where "socialism" was simply a form of religious fanaticism.[78] Like more general criticisms of working-class movements, Huxley's view of Booth's Army as fanatical explained how people given over to such allegedly anarchic, antisocial impulses could form highly disciplined communities. Their reason suppressed by Army regime, Salvation soldiers were simply the blind adherents of a demagogue.

Such critiques implicitly reinforced the authority of men of science, whose inborn powers enabled them to surpass even the commercial and industrious classes in self-restraint.[79] Huxley staged his whole press campaign as a scientific inquiry, a controlled trial in the allegedly neutral space of a respectable newspaper to disprove (or obviate) Booth's own experiment in human improvement conducted in the streets. Huxley's letters were displays of careful empiricism; but more crucially, they were demonstrations of the ability of men of science to be social leaders, to rise calmly above the sentimental and political fray, and attain a truth that was for the benefit of all.

Weighing the testimony before him, which included published pieces by Army deserters, Huxley assembled evidence that showed Booth's organization to be a soul-saving sweatshop, and the general to be a worse taskmaster than any British employer. But more alarming to Huxley than Booth's exploitation of labor was his misappropriation of private property. On 14 November Booth had announced the establishment of a "charitable trust" for contributions to his reform scheme, in order for the work toward that scheme to be kept separate from other Army work. Records and balance sheets were made available for inspection: "every subscriber and the public generally . . . would have the right to call me or my successors to account before the courts in the event of any mismanagement or malversation of the funds entrusted to

---

[78] Huxley 1891a: 53–7.
[79] See the discussion of gentlemanliness, and Huxley's portraits of Darwin, in Chapter 2.

my care."[80] Huxley, armed with the opinion of "an eminent Chancery Queen's Counsel," raised objections to the legal status of this "trust." No established laws could make Booth accountable for the funds at his disposal, the fair use of which was entirely contingent on his own pledge to the moral and social regeneration of society. Booth's trust, Huxley concluded, was "purely religious."[81] That this spiritual form of consignment was comparable to the terms under which scientific elites possessed knowledge and that the justice, if not the legality, of these terms was also subject to question were matters that Huxley would be forced to consider.

In engaging Booth, Huxley was not on the familiar ground prepared by previous disputes involving science and religion. The Army faithful were saved not by the emotions of dependence and obligation favored by liberal Anglicans, but by a passionate conversion experience and rough music. Nor were they compelled by neo-Malthusian critiques of the primitive nature of enthusiasm for an earthly kingdom. Huxley's criticism of Booth's trust presupposed an adherence to certain laws of economy and behavior and an allegiance to certain legal and financial institutions that, according to Booth and his supporters, only served the wealthy and powerful. Huxley soon found that the press was not the sequestered space of the laboratory. Initially based on the "data" of Booth's book, his examination quickly had to incorporate scores of letters and pamplets, many written by Army members, that were sent him after he cast himself in the role of investigator. His scientific inquiry fell under public scrutiny, and he had to defend himself against charges of building a case on the "flimsy foundation of stories told by 3 or 4 Army deserters."[82] "What is the real state of the case?" he wrote,

> Simply this – that having come to the conclusion, from the perusal of *In Darkest England*, that "General" Booth's colossal scheme...was bad in principle and must produce certain evil consequences, and having warned the public to that effect, I quite unexpectedly found my hands full of evidence that the exact evils predicted had, in fact, already shown themselves on a great scale; and, carefully warning the public to criticise this evidence, I produced a small part of it...[Of] the support, encouragement, and information which I have received from active and sincere members of the Salvation Army...I can make no use, because of the terroristic discipline and systematic espionage which my correspondents tell me is enforced by its chief.[83]

[80] The "Darkest England Trust Deed" (1891) is reprinted in Sandall 1955, 3: 328–30.
[81] *Times*, 22 January 1891.
[82] Greenwood 1891: 133. In his letter to the *Times*, 20 December 1890, Huxley had cited from a book by an Army expatriate, *The New Papacy. Behind the Scenes in the Salvation Army, by an ex-Staff Officer* ([Anonymous] 1889a).
[83] Huxley 1891: 133.

Huxley's defense, in which hypotheses about Army behavior were used to filter evidence about those hypotheses, was unsatisfactory to many because the supporting theories about human nature and the authorities who pronounced them were in dispute. Salvationists answered Huxley's charges of despotism against Booth with tales of the general's love and of their own freedom. The liberal Anglican Farrar used another paper, the *Daily Graphic*, to assail every Huxley letter. While Booth, Farrar wrote, had sacrificed respectable position and friends to labor among the masses, Huxley stood by in ignorance and scoffed: "I venture to doubt whether any of those who have filled whole columns of the newspapers with sneers and insults have taken so much care as I have done to acquaint themselves with the facts."[84] Ben Tillet wrote to the *Times* asking why Booth's system needed Huxley's approval, Huxley being but a "theorist, scientist, and word juggler," and Booth "a practical man and social expert":

> The embruted millions will want a strong hand to control them; there must be no sign of division, but a rigorous firmness and undisputed authority. The millions will obey nothing less, will only be loyal to a towering support and guidance ... Let Professor Huxley stick to the laboratory and protoplasm and allow Booth to lead and fight in the rough battle of life, consummating a religion of humanity.[85]

For the writer Robert Buchanan, Huxley could not be the people's liberator and tribune, for he had already declared himself, in his article "On the Natural Inequality of Men," to be the people's police. Jealous because someone of untutored intellect had accomplished more than the churches, including the church of science, Huxley was simply a person who, by scientific temperament, was "deprived of the true philosophic vision and the real enthusiasm of humanity," someone who, despite the best of intentions, "vindicated centuries of wrongdoing, and upheld the tyrannies of force and convention."[86]

In early January 1891, several months after the release of *Darkest England*, a *Times* editorial declared that "the hot fit has passed" and praised Huxley for leading the critical onslaught.[87] But with receipts for his scheme already at ninety thousand pounds, and translations of his book into eight or more languages pending, Booth himself remarked on "the all but universal favour" with which his *Darkest England* proposals

---

[84] *Daily Graphic*, 30 December 1890, pp. 85–6. Farrar had been one of Huxley's chief allies in the reform of public school and university education in the 1860s and 1870s, see Chapter 3.

[85] *Times*, 23 December 1890.

[86] *Times*, 9 December 1890.

[87] *Times*, 27 December 1890.

had been received.[88] The verdict reached by the public on the social
role of men of science was also unclear. Adopting a more manageable
medium, Huxley collected his *Times* letters and certain accountants' reports and running commentary into the pamphlet *Social Diseases and
Worse Remedies*. He prefaced the work with the claim that this pamphlet, as did all his writings on society and politics, exhibited a single
basic principle, empiricism, and that following this principle steered one
down the middle course between "anarchic individualism" and "regi-
mental socialism." He also added a previously published essay, "The
Struggle for Existence in Human Society," casting the entire controversy
in Malthusian terms and appending his own military campaign: technical education. In June 1891, however, the *Trade Unionist* solicited the
opinion of learned men, Huxley included, on the recent bus strike, and
found the results discouraging: "You men of science, art, law, medicine,
state-craft, religion, what light have you to throw on this particular case,
in which men are claiming freedom from toil for twelve hours out of
twenty-four?" The editor asked whether "the survival of the fittest bus
driver" was to be measured "by the capacity to stick on the box longest?"

> [O]ne would be glad to know what is Professor Huxley's opinion on the
> relation of modern learning and erudition to this labour movement in
> which he sees so grave a menace to the State. How is the ripe and excessive
> sanity stored up in universities and guarded by professors to be turned to
> account by those who need it so sorely? . . . The little experiment at tapping
> all this wisdom . . . is not encouraging. The moral of it is, look not to the
> right hand or to the left, but go straight on.[89]

Conclusion: The Limits of Evolution

Toward the end of 1893, Huxley delivered his last public address, entitled "Evolution and Ethics," to a large Oxford audience. In the lecture,
Huxley posed a sharp division, indeed a frank opposition, between
nature and morality. The former remained a realm of "strenuous
battle . . . in which all combatants fall in turn."[90] Human ethics, however,
was presented as a struggle against Malthusian forces, a process rooted
in self-sacrifice and mutual aid, not self-assertion. Natural laws and
evolutionary processes did not produce social progress, they impeded
it. Huxley's lecture provoked surprise among some, who saw it as an
abandonment of the naturalistic agenda that he and other men of science

[88] William Booth speech, as reported in the *Times*, 26 December 1890.
[89] *Trade Unionist*, 20 June 1891.
[90] Huxley 1893: 49.

had made an integral part of their work over the last half-century.[91] To help clarify his position, Huxley attached a long "Prolegomena" to the published version of the address that appeared the following year. Here he portrayed society as a garden carved out of a larger environment (nature). He claimed that his last thirty years had been spent showing the ways in which this garden was a part and product of that environment; but as he now pointed out, the boundaries of the garden had to be constantly defended. Huxley then proposed the necessity for "some administrative authority" to preside over the garden, eliminating rivals, defending against the encroaching wilderness, and selecting, according to a given ideal, which organisms would survive and flourish. Such selection, he suggested, might proceed on the basis of courage, industry, cooperation, and intelligence. After preparing his readers for what seemed a *scientia ex machina*, however, he declared that it was impossible to discern which portions of the garden were best, that virtues and vices were not hereditary, and that a "strictly scientific administration" was impossible. Even the instinctive basis of struggle in human society, self-assertion, could not be uprooted, because it was necessary in the human battle against nature.[92]

Huxley continued to work over these ideas in a piece that he never finished, "Civil History and Natural History," concluding that, while nature had evolved certain ethical impulses, there was no ethics of evolution.

> The so-called "Ethics of Evolution" rests upon the supposition that social progress like natural historical progress is dependent on the struggle for existence. Therefore it insists that outside the charmed circle of "natural rights" the struggle for existence shall have full play – so long as nobody directly interferes with anybody he not only may but ought to get all he can and keep all he gets...
>
> The first great step towards progress in the Civil History world is the establishment of security of life and property for all, without reference to their adaptability or indeed to anything else than the fact that they are some human being. Consequently when that state has been reached – the struggle for existence of the Natural History world is at an end. Moreover from this point Ethical progress – the Evolution of Ethics – is antagonistic to any exercise of the machinery of Natural History Evolution.[93]

Some scholars have read in these last writings of Huxley a rejection of the "social Darwinism" that was being touted by Spencer and others.

---

[91] For contemporary reviews that present Huxley's lecture as a radical shift, see [Anonymous] 1893, Mivart 1893, and Courtney 1895.

[92] Huxley 1893: 23.

[93] "Civil History and Natural History," unpublished manuscript, HP: 45.42–50.

As Robert Richards has noted, Huxley did account for the emergence of altruistic instincts through a process of group selection, that is, natural selection operating on the level of the community, rather than the individual. But he saw no way to explain, by the same evolutionary mechanism, the conversion of these instincts into ethical norms, applied universally to humankind.[94] For this reason, Huxley has often been viewed as an early exponent of the "naturalistic fallacy," the principle that it is illegitimate to derive moral principles from matters of fact. Some interpreters have gone further, and seen Huxley's ethical works as an embrace of "humanism" and a critique of "modernity," with its assumptions of material and moral progress through the accumulation of scientific knowledge.[95] But while Huxley's position on evolutionary ethics may have had implications for a more general critique of science and society, Huxley never drew out these implications. His campaigns for university reform, still ongoing in the 1890s, called for the establishment of academic specialties in history, religion, and the fine arts, not as "humanities," but on a strict model of scientific disciplines; nor did he ever propose any alternative to an advancing scientific civilization as did many fin-de-siècle "antimoderns."

By contrast, Huxley's Oxford lecture has also been interpreted as entirely consistent in political terms with his other writings of the period. His critique of evolutionary ethics could indeed function as an implicit attack on the radical versions of evolutionary theory being made by labor leaders and land nationalists: if the naturalistic basis for all social and political systems was removed, then the critics of capitalism could claim no scientific legitimacy for their views.[96] But such a dichotomy between nature and morality could serve another political end. If neither nature nor the struggle that pervaded it were good, then science, and the natural laws it prescribed, were absolved of moral responsibility. In his debates with Gladstone, Huxley had attempted to resolve problems of political sovereignty and to control the boundaries of elite and popular in the customary fashion of an argument between learned men. This strategy proved inadequate, however, as other groups entered the public sphere and redefined the debate with their own claims to learning and political capacity. In his efforts to free "the people" from socialistic, evangelical, and Gladstonean reigns of terror, Huxley was brought into contact with a new sort of audience, a unified working class. Through

---

[94] R. Richards 1987: 313–19.

[95] On Huxley as a cultural humanist and critic of modernity, see Paradis and Williams 1989: Introduction.

[96] Huxley's political conservativism has been emphasized by Helfand 1977, Jacyna 1980, and Di Gregorio 1984.

encounters with this public and its claims and expectations, and in the face of challenges to intellectual property and liberal programs of public instruction, the identity of scientific practitioners was undermined. Evolutionary theory had been an important part of a liberating moral agenda for Huxley. To remove the ethics from human struggle was to commit the English workers in his technical education scheme to a war that, unlike Booth's war on poverty and pain, they could not hope to win. But neither could they stand to gain by attacking established scientific elites. As expedient as this solution may have been, however, it was not in keeping with the traditional role of men of science as moral authorities. That distinctly Victorian figure, the man of science, was disappearing, and a new persona, the modern scientist, was being born.

Conclusion

The End of the "Man of Science"

Over the course of his career, Huxley worked to define the Victorian "man of science" through a complex set of discriminating categories embracing gender, gentlemanliness, literature, religion, and the relations of "elite" and "popular." An absence of well-established career patterns and institutitonal structures made for an enormous diversity amongst practitioners. Scientific identity thus depended crucially on the assertion of social and cultural "others" to maintain its coherence and to secure its boundaries. In a variety of ways, Huxley's scientific self was positively constituted *of* these others. The "autonomous" scientific practitioner that allegedly emerged in the Victorian period as a result of professionalization was in fact a carefully wrought image that obscured a host of new social relations between men of science and heads of state, industrialists, publishers, and others. It also concealed substantial cultural borrowings from domesticity, from theology, from literature, and from empire.

Conflations of science with other social practices, such as literature and religion, facilitated friendships and working relations across professional boundaries and consolidated a more general authority of cultural elites. Science and literature were conjoined as symbolic, even fictive, creations of genius and imagination that refashioned material reality (and minds). Science and religion in turn were represented as essential components of biblical criticism and of the discovery of natural order.

The joint efforts of elites in the domain of public instruction, such as the London School Board, provided spectacles of social solidarity while institutionalizing the common culture they produced. By presenting as *universal* a culture that was potentially "an engine of class distinction," as Arnold put it, clerics, men of science, and men of letters could appear together as progressive reformers and representatives of a broad public, and they could prescribe the direction of progress and reform. By so doing, they claimed to raise themselves above social divisions and to help transform a society, divided by class, party, and denomination, into a liberal nation composed of self-determining individuals.

The conflicts between men of science and these other social groups, conflicts for which the Victorian period is famous, have thus been approached here as negotiations within a shared set of concerns. Yet as Huxley's wide-ranging debates show, much effort was expended by men of science in defending their terrain against social intrusions and appropriations. In a variety of ways, men of science worked to *exclude* other groups, including other cultural elites, from the making of scientific knowledge. Certainly these efforts were to a considerable degree successful, for they did, ultimately, establish a separate status for participants in the sciences in the Victorian world. It might appear that as a result of this successful policing of boundaries by new professional practitioners, the approach to scientific identity as a negotiated process ultimately fails. However, although men of science were able to gain a virtual monopoly on the production of scientific knowledge, their success at controlling the meaning of this knowledge and of their own cultural identity was less clear-cut.

Much of the cultural authority that men of science acquired during the period derived precisely from apparent trespasses into their domain and from misuses of their powers. It was because Henrietta Huxley continued to interpret Thomas's science theologically that she was able to revere her husband, to have faith in science, and to compose odes to her husband's genius. Arnold and Kingsley favored the introduction of scientific subjects into school and university curricula because, by enlisting science, they could accomplish the reform of classical literary culture, elevate the status of biblical criticism, and strengthen the Anglican Church. It was not until the 1880s, when labor leaders enrolled the sciences to authenticate socialism, that such cultural borrowings ceased to be mutual and became instead threatening to the scientific identity that had been constructed during the middle decades.

Huxley's identity as a "man of science" had been forged in the several decades after mid-century, before groups that ranged from the artisan to the gentleman and in respectable homes, high- and middle-brow journals, workingmen's lectures, and Royal Commissions on liberal and technical education. He had addressed publics whose members valued science both for its moral and religious gravity and for the material benefits it might bestow. He had contracted with entrepreneurs, mingled and sparred with men of letters and clergymen, and taught working men, professionals, and women. He had resided with increasing comfort in institutions devoted to imperial display, commercial advancement, and public instruction. From such positions, he could present himself as the liberator of the people from the dogmas of priestcraft and the idleness of university dons, and as the gentle critic of the coarser practices of capitalism and imperialism. He could claim with impunity that the practical benefits of industry flowed ultimately from his more lofty theoretical pursuits.

In the debates over social reform in the 1880s and 1890s, however, much less of this public persona could be brought to bear. If Huxley found the labor leaders' press campaigns offensive, it was because he was himself on trial before a jury that was neither middle class nor polite. The fate of these issues and the terms in which the debate over them was conducted were not settled by learned elites or their traditional audiences. The public whom Huxley still claimed to instruct and liberate, the "working men" of old, was not the public addressed by Booth or Gladstone. Nor could Huxley negotiate any consensus among ruling interests or other learned groups about how to represent this public. Far from respecting Huxley's authority, as a man of science, to pronounce on matters of public welfare, the labor leader Ben Tillet actually disqualified him as being narrow, partisan, and even – a *scientist*.

In many respects, the term "scientist" was now an appropriate one for the scientific practitioner in Britain. Its connotations of narrow specialism, though despised by Huxley, would also prove useful in resolving the problems that Huxley had experienced with his public image at the end of his career. By the turn of the century, the sciences and their laboratories were well-established parts of the institutional structure of British universities. Although debates continued about the relative value and appropriateness of individual subjects, the teaching of the sciences in elementary and secondary schools had likewise become a matter of course. The existence of these controlled environments removed many of the institutional stakes from the disputes over liberal education in which Huxley had engaged with men of letters and clergymen. Retreat into these prestigious academic quarters could also shelter practitioners from the kinds of criticism and humiliation that Huxley had experienced in the press in the 1890s: an indiscriminate association with the cruelest practices of capitalism. The "man of science," though substantially different from the "gentleman of science" of the early nineteenth century, was still essentially moral. His persona was fashioned according to the principle that knowledge could produce progress, improvement, and happiness, in short, both the private and the public good. This principle was constantly reasserted in debates between learned groups during the middle decades of the nineteenth century, for it conferred upon men of science and other elites a place of social prominence. The debates over socialism show, however, that by the 1880s and 1890s the moral claims of men of science were highly contested. They could no longer trade (or at least bank) on the social benefits of their knowledge.

Huxley's career as a man of science thus presents a paradox. Owing to his persistent presentation of the man of science as embodying moral authority and social responsibility over the course of his career, Huxley had helped to gain laboratory practitioners a secure and eminent place in academic institutions. At the turn of the century, this place was maintained in part through a *denial* of such authority and responsibility. The mid-century "man of science," characterized by moral conduct and social duty, had passed away. Huxley struggled against this demise. By condemning the narrow technical discourse of the "scientist," and by forcefully intervening in social and political debates, he tried to maintain the moral character of the man of science. Yet he found the social debates of the 1890s, in which he engaged with labor leaders and low churchmen, to be distasteful. At the same time, he confronted journalistic and political critics who told him to return to the laboratory. Ultimately, in his later writings on evolution, Huxley himself began to deconstruct the foundations on which his scientific identity had hitherto rested. Nature, he now claimed, was a field of ceaseless battles that culminated in no higher end. Good no longer issued from obedience to the natural laws prescribed by men of science. In his 1893 address "Evolution and Ethics," Huxley suggested, indeed, that a morally prescriptive science was a fallacy: since nature was neither evil nor good, ethical principles could not be derived from its study. In short, the links between nature, morality, and scientific practice had been resoundingly cut.

Huxley's last writings thus effectively removed the scientific high ground from those who wished to make moral claims, and vice versa. In a maneuver that would have profound implications for the twentieth century, he surrendered the moral commitments that had always been fundamental to his vocation, and embraced what would become a standard distinction between "fact" and "value," between science and its social use. The "purity" of the practitioner had changed. It no longer meant detachment from "material pursuits" or "private interests," a detachment that was supposed to confer social responsibility and to legitimate a variety of social engagements. The virtue of scientific practitioners no longer issued from the social value of their work, but from their *indifference* to all such considerations. Thus, as the sciences grew more institutionally powerful and secure, more embedded in industrial, military, and governmental networks, more consequential in daily life, their practitioners were absolved of social responsibilities and placed beyond moral reproach. Like the dictum "art for art's sake" touted by aesthetes like Oscar Wilde during the same period, the pursuit of knowledge for its own sake, rather than for the social good, would become *the* legitimating strategy and raison d'être of the scientist.

Bibliography

Archival Sources

American Philosophical Society Library, Philadelphia, Pennsylvania. T. H. Huxley Papers.
Cambridge University Library. Charles Darwin Papers.
Education Library, University of London. Education documents.
Greater London Public Record Office. London School Board documents.
Huntington Library, San Marino, California. Frances Power Cobbe Papers.
Imperial College of Science, Technology, and Medicine Archives, London. T. H. Huxley Papers and T. H. Huxley–Henrietta Heathorn Correspondence.
Senate House Library, University of London. Herbert Spencer Papers. William Sharpey Correspondence.

Government Documents

First Report of the Scheme of Education Committee. London, 1871.
London School Board Report, together with Minutes and Evidence, for 1871. London, 1872.
U.K. Parliament. 1864. Report of Her Majesty's Commissioners Appointed to Inquire into the Revenues Management of Certain Colleges and Schools, and the Studies Pursued and Instruction Given therein. *Parliamentary Papers, 1864*. Vols. 20–21.
　　1865. Report from the Select Committee by the House of Lords . . . on the Public Schools Bill. *Parliamentary Papers, 1865*. Vol. 10.
　　1867–8. Report from the Select Committee on Scientific Instruction. *Parliamentary Papers, 1867–8*. Vol. 15.

Report of the Royal Commission Known as the Schools Inquiry Commission. *Parliamentary Papers, 1867–8.* Vol. 28.

Report of a Committee Appointed by the Council for the British Association for the Advancement of Science to Consider the Best Means of Promoting Scientific Education in Schools. *Parliamentary Papers, 1867–8.* Vol. 54.

1871. Report of the Committee of Council on Education, 1870–71. *Parliamentary Papers, 1871.* Vol. 21.

Eighteenth Report of the Science and Art Department of the Committee of Council on Education. *Parliamentary Papers, 1871.* Vol. 24.

1872. Royal Commission on Scientific Instruction and the Advancement of Science, First, Supplementary, and Second Report. *Parliamentary Papers, 1872.* Vol. 25.

Other Primary and Secondary Sources

[Anonymous]. 1863. *A Report. A Sad Case, Recently Tried before the Lord Mayor, Owen versus Huxley.* London: n.p.

1886. The Fear of Mobs. *Spectator* 59: 218–19.

1889a. *The New Papacy: Behind the Scenes in the Salvation Army, by an Ex-Staff Officer.* Toronto: n.p.

1889b. The Physical Force of the Mob. *Spectator* 62: 535–6.

1889c. The Control of Crowds. *Spectator* 62: 756–7.

1893. Professor Huxley's Somersault. *Free Review* 1: 18.

1945. *A Short History of the Imperial College, 1845–1945.* London: William Brown and Co.

Abir-Am, Pnina, and Dorinda Outram, eds. 1987. *Uneasy Careers and Intimate Lives: Women in Science, 1789–1978.* New Brunswick: Rutgers University Press.

Alborn, Timothy L. 1996. The Business of Induction: Industry and Genius in the Language of British Scientific Reform, 1820–1840. *History of Science* 34: 91–121.

Allen, David E. 1978. *The Naturalist in Britain: A Social History.* London: Penguin.

Alter, Peter. 1987. *The Reluctant Patron: Science and the State in Britain, 1850–1920.* Angela Davies, trans. Oxford: Berg.

Altholz, Josef. 1994. *Anatomy of a Controversy: The Debate over Essays and Reviews, 1860–1864.* Aldershot: Scholar Press.

Altick, Richard. 1957. *The English Common Reader.* Chicago: University of Chicago Press.

1979. *The Shows of London.* Cambridge, MA: Harvard University Press.

Appel, Toby. 1987. *The Cuvier-Geoffroy Debate: French Biology in the Decades before Darwin.* New York: Oxford University Press.

Armstrong, W. G. 1888a. The Vague Cry for Technical Education. *Nineteenth Century* 24: 325–33.

1888b. The Cry for Useless Knowledge. *Nineteenth Century* 24: 653–68.

Arnold, Matthew. 1853. Preface to 1st Edition of *Poems.* In Super ed. 1977–85, 1: 1–15.

1862a. On Translating Homer. In Super ed. 1977–85, 2: 97–216.

1862b. The Twice-Revised Code. In Super ed. 1977–85, 3: 212–43.

1862c. The Principle of Examination. In Super ed. 1977–85, 3: 244–46.

1862d. The Code out of Danger. In Super ed. 1977–85, 3: 247–51.

1863–4. A French Eton, or Middle-Class Education and the State. In Super ed. 1977–85, 2: 262–325.

1865a. The Function of Criticism at the Present Time. In Super ed. 1977–85, 4: 258–85.

1865b. On the Literary Influence of Academies. In Super ed. 1977–85, 4: 232–57.

1865c. Preface to *Essays in Criticism, 1st Series*. In Super ed. 1977–85, 4: 286–90.

1867. Culture and Its Enemies. *Cornhill Magazine* 16: 36–53.

1868a. *Culture and Anarchy*. In Super ed. 1977–85, 6: 85–230.

1868b. *Schools and Universities on the Continent*. In Super ed. 1977–85, 5: 15–328.

1870. *St. Paul and Protestantism*. In Super ed. 1977–85, 7: 1–127.

1873. *Literature and Dogma*. In Super ed. 1977–85, 7: 139–411.

1879. A Speech at Eton. In Super ed. 1977–85, 9: 20–35.

1882. Literature and Science. In Super ed. 1960–75, 10: 53–73.

Ashton, Rosemary. 1996. *George Eliot: A Life*. London: Allen Lane.

Backstrom, P. N. 1974. *Christian Socialism and Co-operation in Victorian England: Edward Vansittart Neale and the Co-operative Movement*. London: Croom Helm.

Bailey, Victor. 1977. Salvation Army Riots, the "Skeleton Army" and Legal Authority in Provincial Towns. In A. P. Donajgrodzki, ed. *Social Control in Nineteenth-Century Britain*. London: Croom Helm, pp. 231–53.

Barker-Benfield, G. J. 1992. *The Culture of Sensibility: Sex and Society in Eighteenth-Century Britain*. Chicago: University of Chicago Press.

Barlow, Nora, ed. 1958. *The Autobiography of Charles Darwin*. London: Collins.

Barr, Alan, ed. 1997a. *Thomas Henry Huxley's Place in Science and Letters: Centenary Essays*. Athens, GA: University of Georgia Press.

1997b. *The Major Prose of Thomas Henry Huxley*. Athens, GA: University of Georgia Press.

Barton, Ruth. 1983. Evolution: The Whitworth Gun in Huxley's War for the Liberation of Science from Theology. In D. R. Oldroyd and I. Langham, eds. *The Wider Domain of Evolutionary Thought*. Dordrecht: Reidel, pp. 261–86.

1987. John Tyndall, Pantheist. *Osiris* 3: 111–34.

1990. "An Influential Set of Chaps": The X-Club and Royal Society Politics, 1864–85. *British Journal for the History of Science* 23: 53–81.

1998. "Huxley, Lubbock, and Half a Dozen Others": Professionals and Gentlemen in the Formation of the X-Club, 1851–1864. *Isis* 89: 410–44.

Battersby, Christine. 1989. *Gender and Genius: Towards a Feminist Aesthetics*. London: Women's Press.

Becker, B. H. 1874. *Scientific London*. London: King.

Beer, Gillian. 1983. *Darwin's Plots: Evolutionary Narrative in Darwin, George Eliot, and Nineteenth-Century Fiction.* London: Routledge and Kegan Paul.

——— 1996. Travelling the Other Way. In N. Jardine, J. Secord, and E. Spary, eds., pp. 322–37.

Benson, Keith. 1991. From Museum Research to Laboratory Research: The Transformation of Natural History into Academic Biology. In R. Rainger, K. Benson, and J. Maienschein, eds. *The American Development of Biology.* New Brunswick: Rutgers University Press, pp. 49–83.

Berman, Morris. 1978. *Social Change and Scientific Organization: The Royal Institution, 1799–1844.* Ithaca: Cornell University Press.

Bernardin de St. Pierre, Jacques Henri. 1828. *Paul and Virginia* and *The Indian Cottage.* London: J. F. Dove.

Bibby, Cyril. 1959. *T. H. Huxley: Scientist, Humanist, Educator.* London: Watts.

Blinderman, Charles. 1970. The Great Bone Case. *Perspectives in Biology and Medicine* 14: 370–93.

Block, E. 1986. T. H. Huxley's Rhetoric and the Popularization of Victorian Scientific Ideas, 1854–1874. *Victorian Studies* 29: 363–86.

Blunt, Alison. 1994. *Travel, Gender, and Imperialism: Mary Kingsley and West Africa.* New York: The Guilford Press.

Booth, Catherine. 1883. *The Salvation Army in Relation to the Church and State.* London: S. W. Partridge and Co.

Booth, William. 1890. *In Darkest England and the Way Out.* London: William Burgess.

Bowlby, John. 1990. *Charles Darwin: A Biography.* London: Hutchinson.

Branca, Patricia. 1974. Image and Reality: The Myth of the Idle Victorian Woman. In M. Hartman, ed. *Clio's Consciousness Raised: New Perspectives on the History of Women.* New York: Harper and Row, pp. 179–91.

Brantlinger, Patrick. 1997. Thomas Henry Huxley and the Imperial Archive. In Barr, ed., 1997a, pp. 259–76.

——— ed. 1989. *Energy and Entropy: Science and Culture in Victorian Britain.* Bloomington: Indiana University Press.

Briggs, Asa. 1960. The Language of "Class" in Early Nineteenth-Century England. In *The Collected Essays of Asa Briggs*, vol. 1. Brighton: Harvester, 1985, pp. 3–33.

——— 1979. The Language of "Mass" and "Masses" in Nineteenth-Century England. In D. E. Martin and D. Rubinstein, eds. *Ideology and the Labour Movement.* London: Croom Helm, pp. 62–83.

——— 1981. The Salvation Army in Sussex, 1883–1892. In M. J. Kitch, ed. *Studies in Sussex Church History.* London: Leopard's Head Press, pp. 189–208.

——— 1990. *Victorian People.* London: Penguin.

Brock, W. H., and R. M. MacLeod. 1976. The Scientists' Declaration: Reflections on Science and Belief in the Wake of *Essays and Reviews*, 1864–5. *British Journal for the History of Science* 9: 39–66.

Brooke, John. 1991. *Science and Religion: Some Historical Perspectives.* Cambridge: Cambridge University Press.

Brown, A. W. 1947. *The Metaphysical Society.* New York: Columbia University Press.

Browne, Janet. 1992. A Science of Empire: British Biogeography before Darwin. *Revue de l'histoire des sciences modernes et contemporaines* 45: 453–75.

——— 1995. *Charles Darwin: Voyaging*. New York: Knopf.

——— 1998. I Could Have Retched All Night: Darwin's Body. In C. Lawrence and S. Shapin, eds. *Science Incarnate: Historical Embodiments of Natural Knowledge*. Chicago: University of Chicago Press, pp. 240–87.

Burkhardt, Frederick, and Sydney Smith, eds. 1985–2001. *The Correspondence of Charles Darwin*. 12 vols. Cambridge: Cambridge University Press.

Burrow, John W. 1966. *Evolution and Society: A Study in Victorian Social Theory*. Cambridge: Cambridge University Press.

Cannon, Susan F. 1978. *Science in Culture: The Early Victorian Period*. New York: Science History Publications.

Cantor, Geoffroy. 1991. *Michael Faraday: Sandamanian and Scientist*. Basingstoke: Macmillan.

Cardwell, D. S. L. 1972. *The Organisation of Science in England*. London: Heinemann.

Carlyle, Thomas. 1827. Jean Paul Friedrich Richter. *Edinburgh Review* 46: 176–94.

——— 1829. The Life of Heyne. *Foreign Review and Continental Miscellany* 2: 351–83.

——— 1831. Characteristics. *Edinburgh Review* 54: 351–83.

Caron, J. A. 1988. Biology in the Life Sciences: A Historiographical Contribution. *History of Science* 26: 223–68.

Carpenter, William B. 1874. *Principles of Mental Physiology*. London: Henry S. King and Co.

Certeau, Michel de. 1984. *The Practice of Everyday Life*. Steven Rendell, trans. Berkeley: University of California Press.

Chadwick, Owen. 1975. *The Secularization of the European Mind in the Nineteenth Century*. Cambridge: Cambridge University Press.

Chambers, Robert, 1844. *Vestiges of the Natural History of Creation*. In J. Secord, ed. 1994. *Vestiges of the Natural History of Creation and Other Evolutionary Writings*. Chicago: University of Chicago Press.

Chartier, Roger. 1987. *The Cultural Uses of Print in Early Modern France*. Lydia Cochrane, trans. Princeton: Princeton University Press.

Christie, John, and Sally Shuttleworth, eds. 1989. *Nature Transfigured: Science and Literature, 1700–1900*. Manchester: Manchester University Press.

Clark, Anna. 1995. *The Struggle for the Breeches: Gender and the Making of the British Working Class*. London: Rivers Oram Press.

Clark, John, and Thomas Hughes, eds. 1890. *The Life and Letters of the Reverend Adam Sedgwick*. 2 vols. Cambridge: Cambridge University Press.

Clarke, Norma. 1991. Strenuous Idleness: Thomas Carlyle and the Man of Letters as Hero. In Roper and Tosh, eds., pp. 125–43.

Clarke, Peter. 1999. *A Question of Leadership, Gladstone to Blair*. 2nd ed. London: Penguin.

Clifford, William. 1877. The Ethics of Belief. *Contemporary Review* 29: 289–309.

Cobbe, Frances Power. 1894. *Life of Frances Power Cobbe*. 2 vols. London: Richard Bentley and Son.

Cockshut, A. O. J. 1964. *The Unbelievers: English Agnostic Thought, 1840–1890.* London: Collins.

Colenso, John William. 1862–79. *The Pentateuch and Book of Joshua Critically Examined.* 7 vols. London: Longman and Co.

Collier, Richard. 1965. *The General Next to God: The Story of William Booth and the Salvation Army.* London: Collins.

Collini, Stefan. 1989. Manly Fellows: Fawcett, Stephen and the Liberal Temper. In L. Goldman, ed. *The Blind Victorian: Henry Fawcett and British Liberalism.* Cambridge: Cambridge University Press, pp. 40–59.

——— 1991. *Public Moralists: Political Thought and Intellectual Life in Britain, 1850–1930.* Oxford: Clarendon Press.

Combe, George. 1852. On Secular Education. *Westminster Review* 58: 1–32.

Connell, W. F. 1950. *The Educational Thought and Influence of Matthew Arnold.* London: Routledge and Kegan Paul.

Cooter, Roger. 1984. *The Cultural Meaning of Popular Science: Phrenology and the Organization of Consent in Nineteenth-Century Britain.* Cambridge: Cambridge University Press.

——— and Stephen Pumphrey. 1994. Separate Spheres and Public Places: Reflections on the History of Science Popularization and Science in Popular Culture. *History of Science* 32: 237–67.

Coulling, Sydney. 1974. *Matthew Arnold and His Critics: A Study of Arnold's Controversies.* Athens, OH: Ohio University Press.

Courtney, W. L. 1895. Professor Huxley as a Philosopher. *Fortnightly Review* 64: 317–22.

Cowell, F. R. 1975. *The Athenaeum: Club and Social Life in London, 1824–1974.* London: Heineman.

Cowling, Maurice. 1967. *1867, Disraeli, Gladstone and Revolution: The Passing of the Second Reform Bill.* Cambridge: Cambridge University Press.

Crook, Paul. 1994. *Darwinism, War and History: The Debate over the Biology of War from the "Origin of Species" to the First World War.* Cambridge: Cambridge University Press.

Cross, Nigel. 1985. *The Common Writer: Life in Nineteenth-Century Grub Street.* Cambridge: Cambridge University Press.

Curle, Richard. 1954. *The Ray Society: A Bibliographic History.* London: Ray Society.

Dale, Peter. 1986. George Lewes's Scientific Aesthetic: Restructuring the Ideology of the Symbol. In Levine, ed., pp. 92–116.

——— 1989. *In Pursuit of a Scientific Culture: Science, Art and Society in the Victorian Age.* Madison: University of Wisconsin Press.

Darwin, Angela, and Adrian Desmond, eds. Forthcoming. *The Thomas Henry Huxley Family Correspondence.* 4 vols. Chicago: University of Chicago Press.

Darwin, Francis, ed., 1887. *The Life and Letters of Charles Darwin.* 3 vols. London: John Murray.

Daston, Lorraine. 1992. The Naturalized Female Intellect. *Science in Context* 5: 209–35.

——— and Peter Galison. 1992. The Image of Objectivity. *Representations* 40: 81–128.

David, Deirdre. 1987. *Intellectual Women and Victorian Patriarchy.* London: Macmillan.

Davidoff, Leonore, and Catherine Hall. 1987. *Family Fortunes: Men and Women of the English Middle-Class, 1780–1850.* Chicago: University of Chicago Press.

Dear, Peter, ed. 1991. *The Literary Structure of Scientific Argument.* Philadelphia: University of Pennsylvania Press.

Desmond, Adrian. 1982. *Archetypes and Ancestors: Palaeontology in Victorian London, 1850–1875.* Chicago: University of Chicago Press.

——— 1987. Artisan Resistance and Evolution in Britain, 1819–1848. *Osiris* 3: 77–110.

——— 1989. *The Politics of Evolution: Morphology, Medicine, and Reform in Radical London.* Chicago: University of Chicago Press.

——— 1998. *Thomas Huxley: From Devil's Disciple to Evolution's High Priest.* London: Penguin.

——— 2001. Redefining the X Axis: "Professionals," "Amateurs" and the Making of Mid-Victorian Biology – A Progress Report. *Journal of the History of Biology* 34: 3–50.

Di Gregorio, Mario. 1984. *T. H. Huxley's Place in Natural Science.* New Haven: Yale University Press.

Dockrill, D. W. 1971. T. H. Huxley and the Meaning of "Agnosticism." *Theology* 74: 461–77.

Drayton, Richard. 2000. *Nature's Government: Science, Imperial Britain, and the "Improvement" of the World.* New Haven: Yale University Press.

Drummond, James, and C. B. Upton, eds. 1902. *The Life and Letters of James Martineau.* 2 vols. London: James Nisbet and Co.

Durant, J. 1979. Scientific Naturalism and Social Reform in the Thought of Alfred Russel Wallace. *British Journal for the History of Science* 12: 31–58.

Eagleton, Terry. 1984. *The Function of Criticism: From the Spectator to Post-structuralism.* London: Verso.

Eliot, George. 1856a. The Natural History of German Life. *Westminster Review* 66: 51–79.

——— 1856b. Silly Novels by Lady Novelists. *Westminster Review* 64: 442–61.

Ellegård, Alvar. 1958. *Darwin and the General Reader: The Reception of Darwin's Theory of Evolution in the British Periodical Press, 1859–1872.* Gøteborg: Gøteborg University Press.

Engel, Arthur. 1983. *From Clergyman to Don: The Rise of the Academic Profession in Nineteenth-Century Oxford.* New York: Oxford University Press.

Ervine, St. John. 1934. *God's Soldier: General William Booth.* 2 vols. London: Heinemann.

Eve, A. S., and C. H. Creasey. 1945. *Life and Work of John Tyndall.* London: Macmillan.

Fish, Stanley. 1980. *Is There a Text in This Class?* Cambridge, MA: Harvard University Press.

Forbes, Edward. 1855. *Literary Papers by the Late Professor Edward Forbes, F. R. S., Selected from His Writings in the* Literary Gazette. London: Lovell Reeve.

Forgan, Sophie. 1994. The Architecture of Display: Museums, Universities and Objects in Nineteeth-Century Britain. *History of Science* 32: 139–62.

and Graeme Gooday. 1994. "A Fungoid Assemblage of Buildings": Diversity and Adversity in the Development of College Architecture and Scientific Education in Nineteenth-Century South Kensington. *History of Universities* 13: 153–92.

1996. Constructing South Kensington: The Buildings and Politics of T. H. Huxley's Working Environments. *British Journal for the History of Science,* 29: 435–68.

Foster, Michael, and E. Ray Lankester, eds. 1898–1902. *The Scientific Memoirs of Thomas Henry Huxley.* 5 vols. and supplement. London: Macmillan.

Galton, Francis. 1869. *Hereditary Genius.* London: Macmillan.

1874. *English Men of Science: Their Nature and Nurture.* London: Macmillan.

Gardiner, B. G. 1993. Edward Forbes, Richard Owen and the Red Lions. *Archives of Natural History* 20: 349–72.

Gates, Barbara, and Ann Shteir, eds. 1997. *Natural Eloquence: Women Reinscribe Science.* Madison: University of Wisconsin Press.

Gay, Hannah. 1997. East End, West End: Science Education, Culture and Class in Mid-Victorian London. *Canadian Journal of History* 32: 153–83.

and John W. Gay. 1997. Brothers in Science: Science and Fraternal Culture in Nineteenth-Century Britain. *History of Science* 35: 425–53.

Geison, Gerald L. 1978. *Michael Foster and the Cambridge School of Physiology.* Princeton: Princeton University Press.

Gerard, Alexander. 1774. *An Essay on Genius.* London: Strahan and Cadell.

Gilbert, A. D. 1976. *Religion and Society in Industrial England, 1740–1914.* London: Longman.

Gilley, Sheridan, and Ann Loades. 1981. Thomas Henry Huxley: The War between Science and Religion. *The Journal of Religion* 61: 285–308.

Gillispie, Charles. 1951. *Genesis and Geology.* Cambridge, MA: Harvard University Press.

Girouard, Mark. 1981. *The Return to Camelot: Chivalry and the English Gentleman.* New Haven: Yale University Press.

Gladstone, William. 1877a. On the Influence of Authority in Matters of Opinion. *Nineteenth Century* 1: 2–22.

1877b. Rejoinder on Authority in Matters of Opinion. *Nineteenth Century* 1: 902–26.

Goldgar, Anne. 1995. *Impolite Learning: Conduct and Community in the Republic of Letters, 1680–1750.* New Haven: Yale University Press.

Goldstein, Jan. 1987. *Console and Classify: The French Psychiatric Profession in the Nineteenth-Century.* Cambridge: Cambridge University Press.

1994. The Uses of Male Hysteria: Medical and Literary Discourse in Nineteenth-Century France. In A. La Berge and M. Feingold, eds. *French Medical Culture in the Nineteenth Century.* Amsterdam: Rodopi, pp. 210–47.

Gooday, Graeme. 1991. Nature in the Laboratory: Domestication and Discipline with the Microscope in Victorian Life Science. *British Journal for the History of Science* 24: 307–41.

1997. Instrumentation and Interpretation: Managing and Representing the Working Environments of Victorian Experimental Science. In Lightman, ed., pp. 409–37.

Gould, Paula. 1997. Women and the Culture of University Physics in Late Nineteenth-Century Cambridge. *British Journal for the History of Science* 30: 127–49.

Gray, Robert. 1981. *The Aristocracy of Labour in Nineteenth-Century Britain, c. 1850–1900*. London: Macmillan.

Greenblatt, Stephen. 1980. *Renaissance Self-Fashioning*. Chicago: University of Chicago Press.

Greenwood, H. J. 1891. *General Booth and His Critics*. London: n.p.

Gross, John. 1969. *The Rise and Fall of the Man of Letters: Aspects of English Literary Life since 1800*. London: Penguin.

Haight, Gordon. 1940. *George Eliot and John Chapman*. New Haven: Yale University Press.

    ed. 1954–78. *The George Eliot Letters*. 9 vols. New Haven: Yale University Press.

Harrison, Frederic. 1867. Culture: A Dialogue. *Fortnightly Review* 8: 603–14.

Harrison, J. F. C. 1954. *A History of the Working Men's College, 1854–1954*. London: Routledge and Kegan Paul.

Harte, Negley. 1986. *The University of London, 1836–1986*. London: Athlone Press.

Hays, J. N. 1974. Science in the City: The London Institution, 1818–40. *British Journal for the History of Science* 7: 146–62.

    1983. The London Lecturing Empire, 1800–50. In Inkster and Morrell, eds., pp. 91–119.

Helfand, Michael. 1977. T. H. Huxley's "Evolution and Ethics": The Politics of Evolution and the Evolution of Politics. *Victorian Studies* 20: 159–77.

Heyck, T. W. 1982. *The Transformation of Intellectual Life in Victorian England*. London: Croom Helm.

    1987. The Idea of a University in Britain, 1870–1970. *History of European Ideas* 8: 208–20.

Hilton, Boyd. 1988. *The Age of Atonement: The Influence of Evangelicalism on Social and Economic Thought, 1785–1865*. Oxford: Clarendon Press.

    1989. Manliness, Masculinity and the Mid-Victorian Temperament. In L. Goldman, ed. *The Blind Victorian: Henry Fawcett and British Liberalism*. Cambridge: Cambridge University Press, pp. 60–70.

Hobsbawm, E. J. 1964. *Labouring Men*. New York: Basic Books.

    1976. *Primitive Rebels*. New York: Harvester.

    1984. *Worlds of Labour: Further Studies in the History of Labour*. London: Weidenfeld and Nicolson.

Honan, Park. 1981. *Matthew Arnold: A Life*. New York: McGraw Hill.

Hoppen, K. Theodore. 1998. *The Mid-Victorian Generation, 1846–1886*. Oxford: Clarendon Press.

Houghton, Walter. 1957. *The Victorian Frame of Mind*. New Haven: Yale University Press.

Howsam, Leslie. 2000. An Experiment with Science for the Nineteenth-Century Book Trade: The International Scientific Series. *British Journal for the History of Science* 33: 187–207.

Hull, David. 1973. *Darwin and His Critics*. Cambridge, MA: Harvard University Press.

Hutton, R. H. 1870. Pope Huxley. *Spectator*, 29 January 1970, pp. 135–6.

1885. The Metaphysical Society: A Reminiscence. *Nineteenth Century* 18: 177–96.

Huxley, Aldous. 1932. *T. H. Huxley as a Man of Letters*. London: Macmillan and Co.

Huxley, Henrietta A. 1913. *Poems of Henrietta A. Huxley with Three of Thomas H. Huxley*. London: Duckworth and Co.

Huxley, Julian, ed. 1936. *T. H. Huxley's Diary of the Voyage of H. M. S. Rattlesnake*. Garden City, NY: Doubleday, Doran, and Co.

Huxley, Leonard, ed. 1900. *Life and Letters of Thomas Henry Huxley*. 2 vols. London: Macmillan.

Huxley, T. H. 1849. On the Anatomy and Affinities of the Family of the Medusae. [*Philosophical Transactions of the Royal Society* 2: 413.] In Foster and Lankester, eds., 1898–1902, 1: 9–32.

1853. On the Morphology of the Cephalous Mollusca. In Foster and Lankester, eds., 1898–1902, 1: 152–93.

1854a. Science at Sea. *Westminster Review* 61: 98–119.

1854b. The Vestiges of Creation. [*British and Foreign Medico-Chirurgical Review* 26: 425–39.] In Foster and Lankester, eds., 1898–1902, 5: 1–19.

1854c. On the Educational Value of the Natural History Sciences. [Address delivered at St. Martin's Hall.] In T. H. Huxley, 1893–4, 3: 38–65.

1854–7. Contemporary Literature: Science. *Westminster Review* 61: 254–70, 580–95; 62: 242–56, 572–80; 63: 239–53, 558–63; 64: 240–55, 565–74; 65: 261–71; 67: 279–88.

1855. On Certain Zoological Arguments Commonly Adduced in Favour of the Hypothesis of the Progressive Development of Animal Life in Time. [Lecture at the Royal Institution.] In Foster and Lankester, eds., 1898–1902, 1: 300–4.

1856a. Owen and Rymer Jones on Comparative Anatomy. *British and Foreign Medico-Chirurgical Review* 35: 1–27.

1856b. On Natural History, as Knowledge, Discipline, and Power. [Lecture at the Royal Institution.] In Foster and Lankester, eds., 1898–1902, 1:305–14.

1859a. *The Oceanic Hydrozoa*. London: Ray Society.

1859b. Time and Life: Mr. Darwin's "Origin of Species." *Macmillan's Magazine* 1: 142–8.

1859c. Darwin on the Origin of Species. [*The Times*, 26 December 1859.] In T. H. Huxley, 1893–4, 2: 1–21.

1859d. Science and Religion. *The Builder*, 15 January 1859, pp. 35–6.

1860. The Origin of Species. [*Westminster Review*, n.s. 17: 541–70.] In T. H. Huxley, 1893–4, 2: 22–79.

1861. On the Zoological Relations of Man with the Lower Animals. *Natural History Review* 1: 68–84.

1862. *On Our Knowledge of the Causes of the Phenomena of Organic Nature.* [London: Hardwicke.] In T. H. Huxley, 1893–4, 2: 303–475.

1863. *Evidence as to Man's Place in Nature.* [London: Williams and Norgate.] In T. H. Huxley, 1893–4, 7: 1–208.

1864. Science and Church Policy. *The Reader* 4: 821.

1865. Emancipation – Black and White. [*The Reader* 5: 561–2.] In T. H. Huxley, 1893–4, 3: 66–75.

1866. On the Adviseableness of Improving Natural Knowledge. [*Fortnightly Review* 3: 626–37.] In T. H. Huxley, 1893–4, 1: 18–41.

1868a. A Liberal Education and Where to Find It. [*Macmillan's Magazine* 17: 367–78.] In T. H. Huxley, 1893–4, 3: 76–110.

1868b. On the Physical Basis of Life. [*Fortnightly Review* 5: 129–45.] In T. H. Huxley, 1893–4, 1: 130–65.

1869. Scientific Education: Notes of an After-Dinner Speech. [*Macmillan's Magazine* 20: 177–84.] In T. H. Huxley, 1893–4, 3: 111–33.

1870a. On Descartes' "Discourse Touching the Method of Using One's Reason Rightly and of Seeking Scientific Truth." [*Macmillan's Magazine* 22: 69–80.] In T. H. Huxley, 1893–4, 1: 166–98.

1870b. The School Boards: What They Can Do and What They May Do. [*Contemporary Review* 16: 1–15.] In T. H. Huxley, 1893–4, 3: 374–403.

1871a. Mr. Darwin's Critics. [*Contemporary Review* 18: 443–76.] In T. H. Huxley, 1893–4, 2: 120–86.

1871b. Administrative Nihilism [*Fortnightly Review* 10: 525–43.] In T. H. Huxley, 1893–4, 1: 251–89.

1874. Universities: Actual and Ideal. [*Contemporary Review* 23: 657–79.] In T. H. Huxley, 1893–4, 3: 189–234.

1876. On the Study of Biology. In T. H. Huxley, 1893–4, 3: 262–93.

1877a. *Physiography: An Introduction to the Study of Nature.* London: Macmillan.

1877b. Technical Education. [*Fortnightly Review* 23: 48–58.] In T. H. Huxley, 1893–4, 3: 404–26.

1880. Science and Culture [Address at Sir Josiah Mason's College, Birmingham.] T. H. Huxley, 1893–4, 3: 134–59.

1882a. Charles Darwin. *Nature* 25: 597.

1882b. On Science and Art in Relation to Education. [Address at the Liverpool Institution.] In T. H. Huxley, 1893–4, 3: 160–88.

1885a. The Darwin Memorial. [Address at the British Museum of Natural History.] In T. H. Huxley, 1893–4, 2: 248–52.

1885b. The Interpreters of Genesis and the Interpreters of Nature. [*Nineteenth Century* 18: 849–60.] In T. H. Huxley, 1893–4, 4: 139–63.

1886a. Mr. Gladstone and Genesis. [*Nineteenth Century* 19: 191–205.] In T. H. Huxley, 1893–4, 4: 164–200.

1886b. The Evolution of Theology: An Anthropological Study. [*Nineteenth Century* 19: 346–65.] In T. H. Huxley, 1893–4, 4: 485–506.

1887a. Address on Behalf of the National Association for the Promotion of Technical Education. In T. H. Huxley, 1893–4, 3: 427–51.

1887b. An Olive Branch from America. *Nineteenth Century* 22: 620–4.

1887c. The Progress of Science. In T. H. Huxley, 1893–4, 1: 42–129.

1888a. Obituary Notice: Charles Robert Darwin. [*Proceedings of the Royal Society of London* 44: i–xxiv.] In T. H. Huxley, 1893–4, 2: 253–302.

1888b. The Struggle for Existence in Human Society. [*Nineteenth Century* 23: 161–80.] In T. H. Huxley, 1893–4, 9: 195–236.

1888c. "How to Become an Orator." *Pall Mall Gazette*, 24 October 1888, pp. 1–2.

1890a. The Keepers of the Herd of Swine. [*Nineteenth Century* 28: 366–92.] In T. H. Huxley, 1893–4, 5: 366–92.

1890b. The Natural Inequality of Men. [*Nineteenth Century* 27: 1–23.] In T. H. Huxley, 1893–4, 1: 290–335.

1890c. Natural Rights and Political Rights. [*Nineteenth Century* 27: 173–95.] In T. H. Huxley, 1893–4, 1: 336–82.

1890d. Capital, the Mother of Labour. [*Nineteenth Century* 27: 513–32.] In T. H. Huxley, 1893–4, 9: 147–87.

1890e. Government: Anarchy or Regimentation. [*Nineteenth Century* 27: 843–66.] In T. H. Huxley, 1893–4, 1: 383–430.

1890f. Autobiography. In T. H. Huxley, 1893–4, 1: 1–17.

1891a. *Social Diseases and Worse Remedies*. London: Macmillan.

1891b. Illustrations of Mr. Gladstone's Controversial Methods. [*Nineteenth Century* 29: 455–67.] In T. H. Huxley, 1893–4, 5: 393–419.

1893. Evolution and Ethics. [Romanes Lecture, Oxford University.] In T. H. Huxley, 1893–4, 9: 46–116.

1893–4. *Collected Essays*. 9 vols. London: Macmillan.

1894. Prolegomena. In T. H. Huxley, 1893–4, 9: 1–43.

Hyndman, Henry M. 1884. *Socialism and Slavery*. London: The Modern Press.

Inkster, Ian. 1976. The Social Context of an Educational Movement: A Revisionist Approach to the English Mechanics' Institutes, 1820–1850. *Oxford Review of Education* 2: 277–307.

and Jack Morrell, eds. 1983. *Metropolis and Province: Science in British Culture, 1780–1850*. London: Hutchinson.

Irvine, William. 1959. *Apes, Angels, and Victorians: Darwin, Huxley, and Evolution*. Cleveland: Meridian.

Jacyna, L. S. 1980. Science and Social Order in the Thought of A. J. Balfour. *Isis* 71: 11–34.

Jardine, Nicholas. 1992. The Laboratory Revolution in Medicine as Rhetorical and Aesthetic Accomplishment. In A. Cunningham and P. Williams, eds. *The Laboratory Revolution in Medicine*. Cambridge: Cambridge University Press, pp. 304–23.

James Secord, and Emma Spary, eds. 1996. *Cultures of Natural History*. Cambridge: Cambridge University Press.

Jensen, J. Vernon. 1988. Return to the Wilberforce-Huxley Debate. *British Journal for the History of Science* 21: 161–79.

1991. *Thomas Henry Huxley: Communicating for Science*. Newark: University of Delaware Press.

Johns, Adrian. 1996. The Physiology of Reading in Restoration England. In
    J. Raven, H. Small, and N. Tadmor, eds., pp. 138–64.
Johnson, R. 1979. "Really Useful Knowledge:" Radical Education and
    Working-Class Culture, 1790–1848. In J. Clarke, C. Critcher, and
    R. Johnson, eds. *Working-Class Culture*. London: Hutchinson, pp. 75–102.
Jones, A. 1996. *Powers of the Press: Newspapers, Power and the Public in
    Nineteenth-Century Fiction*. Aldershot: Scholar Press.
Jones, G. Stedman. 1983. *Languages of Class: Studies in English Working Class
    History, 1832–1982*. Cambridge: Cambridge University Press.
Jordanova, Ludmilla. 1989a. *Sexual Visions: Images of Gender in Science and
    Medicine between the Eighteenth and Twentieth Centuries*. London:
    Harvester.
    1989b. Objects of Knowledge: A Historical Perspective on Museums. In
    P. Vergo, ed. *The New Museology*. London: Reaktion Books, pp. 22–40.
    1993. Gender and the Historiography of Science. *British Journal for the
    History of Science* 26: 469–83.
Joyce, Patrick. 1991. *Visions of the People: Industrial England and the Question of
    Class, 1848–1914*. Cambridge: Cambridge University Press.
Kargon, Robert. 1977. *Science in Victorian Manchester: Enterprise and Expertise*.
    Baltimore: Johns Hopkins University Press.
Kent, Christopher. 1969. Higher Journalism and the Mid-Victorian Clerisy.
    Victorian Studies 13: 181–98.
    1978. *Brains and Numbers: Elitism, Comtism, and Democracy in Mid-Victorian
    England*. Toronto: University of Toronto Press.
Kevles, Daniel. 1985. *In the Name of Eugenics: Genetics and the Uses of Human
    Heredity*. New York: Knopf.
Kingsley, Charles. 1863. *The Water Babies*. London: Macmillan and Co.
Knight, David. 1996. *Science as Power: Humphry Davy*. Cambridge: Cambridge
    University Press.
Kropotkin, Peter. 1888a. The Breakdown of Our Industrial System. *Nineteenth
    Century* 23: 497–516.
    1888b. The Coming Reign of Plenty. *Nineteenth Century* 23: 817–37.
    1888c. The Industrial Village of the Future. *Nineteenth Century* 24: 513–30.
LaCapra, Dominic, and Jeffrey Kaplan, eds. 1987. *Modern European Intellectual
    History: Reappraisals and New Perspectives*. Ithaca: Cornell University
    Press.
Lang, Cecil Y., ed. 1996–2001. *The Letters of Matthew Arnold*. 5 vols.
    Charlottesville: University Press of Virginia.
Lang, Jennifer. 1978. *The City and Guilds of London Institute Centenary, 1878–
    1978: An Historical Commentary*. London: City and Guilds of London
    Institute.
Lankester, E. Ray. 1871. Instruction to Science Teachers at South Kensington.
    Nature 4: 362.
Laurent, J. 1984. Science, Society, and Politics in Late Nineteenth-Century
    England: A Further Look at Mechanics' Institutes. *Social Studies of Science*
    14: 585–619.

Layton, David. 1973. *Science for the People: The Origins of the School Science Curriculum in England.* London: Allen and Unwin.

Levine, George, ed. 1987. *One Culture: Essays in Science and Literature.* Madison: University of Wisconsin Press.

Lewes, George H. 1842a. Hegel's Aesthetics: Philosophy of Art. *British and Foreign Review* 8: 1–49.

1842b. The Errors and Abuses of English Criticism. *Westminster Review* 38: 466–86.

1853. *Comte's Philosophy of the Sciences.* London: George Bell and Sons.

1856. Sea-Side Studies. *Blackwood's Edinburgh Magazine* 80: 184–97, 312–25, 472–85.

1857. New Sea-Side Studies. *Blackwood's Edinburgh Magazine* 81: 667–85; 82: 1–17; 83: 220–40, 345–57, 410–23.

1858. Realism in Art: Recent German Fiction. *Westminster Review* 14: 488–518.

1865. The Principle of Success in Literature. *Fortnightly Review,* vols. 1 and 2.

Lightman, Bernard. 1983. Pope Huxley and the Church Agnostic: The Religion of Science. *Historical Papers of the Canadian Society of History*: 150–63.

1987. *The Origins of Agnosticism: Victorian Unbelief and the Limits of Knowledge.* Baltimore: Johns Hopkins University Press.

1997. "The Voices of Nature": Popularizing Victorian Science. In Lightman, ed., pp. 187–211.

ed. 1997. *Victorian Science in Context.* Chicago: University of Chicago Press.

Livingston, James. 1974. *The Ethics of Belief: An Essay on the Victorian Religious Conscience.* Tallahassee: American Academy of Religion.

1988. Matthew Arnold's Place in the Religious Thought of the Past Century. In C. Machann and F. Burt, eds. *Matthew Arnold in His Time and Ours: Centenary Essays.* Charlottesville: University of Virginia Press, pp. 30–9.

Lockyer, Joseph Norman. 1888. Lord Armstrong on Technical Education. *Nature* 38: 313–14.

1906. *Education and National Progress: Essays and Addresses, 1870–1905.* London: Macmillan and Co.

Lorimer, D. A. 1988. Theoretical Racism in Late Victorian Anthropology, 1870–1900. *Victorian Studies* 31: 405–30.

Lowe, Robert. 1867. Primary and Classical Education. In D. Reeder, ed. *Educating Our Masters.* Leicester: Leicester University Press, 1980, pp. 103–26.

Lowe, Roy. 1987. Structural Change in English Higher Education, 1870–1920. In D. Müller, F. Ringer, and B. Simon, eds., pp. 163–78.

Lubenow, W. C. 1998. *The Cambridge Apostles, 1820–1914: Liberalism, Imagination, and Friendship in British Intellectual and Professional Life.* Cambridge: Cambridge University Press.

Lucas, J. R. 1979. Wilberforce and Huxley: A Legendary Encounter. *Historical Journal* 22: 313–30.

Lyons, Sherrie. 1997. Convincing Men They Are Monkies. In Barr, ed., 1997a, pp. 95–118.

McCarthy, Gerald, ed. 1986. *The Ethics of Belief.* Atlanta: Scholar's Press.

McGeachie, James. 1990. From Parson-Hunter to Eco-Prophet. *History of Science* 28: 429–41.

MacGillivray, John. 1852. *Narrative of the Voyage of the H.M.S. Rattlesnake.* 2 vols. London: T. and W. Boone.

McLeod, Hugh. 1984. *Religion and the Working Class in Nineteenth-Century Britain.* London: Macmillan.

MacLeod, Roy M. 1970a. Science and the Civil List, 1824–1914. *Technology and Society* 6: 47–55.

　　1970b. The X Club: A Social Network of Science in Late-Victorian England. *Notes and Records of the Royal Society* 24: 305–22.

　　1971a. The Royal Society and the Government Grant: Notes on the Administration of Scientific Research, 1849–1914. *Historical Journal* 14: 323–58.

　　1971b. Of Medals and Men: A Reward System in Victorian Science, 1826–1914. *Notes and Records of the Royal Society* 26: 81–108.

　　1977. Education: Scientific and Technical. In G. Sutherland, ed. *Education in Britain.* Dublin: Irish Academic Press, pp. 196–217.

　　1983. Whigs and Savants: Reflections on the Reform Movement in the Royal Society, 1830–48. In I. Inkster and J. Morrell, eds., pp. 55–90.

　　ed. 1988. *Government and Expertise: Specialists, Administrators and Professionals, 1860–1919.* Cambridge: Cambridge University Press.

McRae, Murdo, ed. 1993. *The Literature of Science: Perspectives on Popular Scientific Writing.* Athens, GA: University of Georgia Press.

Mann, Tom. 1896. *How I Became a Socialist.* London: The Modern Press.

Mansel, Henry. 1858. *The Limits of Religious Thought Examined.* London: John Murray.

Martineau, James. 1877. The Influence upon Morality of a Decline in Religious Belief. *Nineteenth Century* 1: 341–5.

Maurice, Frederick D. 1837–8. *The Kingdom of Christ.* 3 vols. London: Darton and Clark.

Max Müller, Friedrick. 1873. *Introduction to the Science of Religion.* London: Longmans, Green, and Co.

Mendelsohn, Everett. 1964. The Emergence of Science as a Profession in Nineteenth-Century Europe. In K. Hill, ed. *The Management of Scientists.* Boston: Beacon Press, pp. 3–48.

Mill, James. 1829. *An Analysis of the Phenomena of the Human Mind.* 2 vols. London: Baldwin and Cradock.

Mill, John Stuart. 1961. *Autobiography.* New York: Columbia University Press.

Mills, Eric L. 1984. A View of Edward Forbes, Naturalist. *Archives of Natural History* 11: 365–93.

Mills, Sara. 1994. Knowledge, Gender, and Empire. In A. Blunt, and G. Rose, eds. *Writing Women and Space: Colonial and Postcolonial Geographies.* New York: The Guilford Press, pp. 29–50.

Mivart, St. George. 1893. Evolution in Professor Huxley. *Nineteenth Century* 34: 198–211.

Moore, James R. 1979. *The Post-Darwinian Controversies: A Study of the Protestant Struggle to Come to Terms with Darwin in Great Britain and America, 1870–1900.* Cambridge: Cambridge University Press.

    1985. Darwin of Down: The Evolutionist as Squarson-Naturalist. In D. Kohn, ed. *The Darwinian Heritage.* Princeton: Princeton University Press, pp. 435–81.

    1990. Theodicy and Society: The Crisis of the Intelligentsia. In R. Helmstadter and B. Lightman, eds. *Victorian Faith in Crisis.* London: Macmillan, pp. 153–86.

Morgan, Marjorie. 1994. *Manners, Morals and Class in England, 1774–1858.* Basingstoke: Macmillan.

Morrell, J. 1971. Individualism and the Structure of British Science in 1830. *Historical Studies in the Physical Sciences* 3: 183–204.

    and A. Thackray. 1981. *Gentlemen of Science: Early Years of the British Association for the Advancement of Science.* Oxford: Clarendon Press.

Morris, William. 1885. *Useful Work and Usless Toil.* London: Socialist League.

Morus, Iwan, Simon Schaffer, and James Secord. 1992. Scientific London. In C. Fox, ed. *London: World City, 1800–1840.* New Haven: Yale University Press, pp. 129–42.

Mullen, Shirley. 1987. *Organized Freethought: The Religion of Unbelief in Victorian England.* New York: Garland.

Müller, Detlef, Fritz Ringer, and Brian Simon, eds. 1987. *The Rise of the Modern Education System: Structural Change and Social Reproduction, 1870–1920.* Cambridge: Cambridge University Press.

Murray, Penelope, ed. 1989. *Genius: The History of an Idea.* Oxford: Oxford University Press.

Myers, Greg. 1989. Science for Women and Children: The Dialogue of Popular Science in the Nineteenth Century. In J. Christie and S. Shuttleworth, eds., pp. 171–200.

Neve, Michael, 1983. Science in a Commercial City: Bristol, 1820–60. In I. Inkster and J. Morrell, eds., pp. 179–204.

Newsome, David. 1961. *Godliness and Good Learning: Four Studies on a Victorian Ideal.* London: Cassell.

Norman, E. R. 1976. *Church and Society in England, 1700–1900.* Oxford: Clarendon Press.

    1985. *The Victorian Christian Socialists.* Cambridge: Cambridge University Press.

Nyhart, Lynn. 1995. *Biology Takes Form: Animal Morphology and the German Universities, 1800–1900.* Chicago: University of Chicago Press.

    1996. Natural History and the "New" Biology. In N. Jardine, J. Secord, and E. Spary, eds., pp., 426–43.

Obelkevich, James. 1976. *Religion and Rural Society: South Lindsey, 1825–1875.* Oxford: Oxford University Press.

Offer, John, ed. 2000. *Herbert Spencer: Critical Assessments.* 4 vols. London: Routledge.

Oppenheim, Janet. 1985. *The Other World: Spiritualism and Psychical Research in England, 1850–1914.* Cambridge: Cambridge University Press.

Outram, Dorinda. 1984. *George Cuvier: Vocation, Science and Authority in Post-Revolutionary France.* Manchester: Manchester University Press.

1987. Before Objectivity: Wives, Patronage, and Cultural Reproduction in Early Nineteenth Century French Science. In P. Abir-Am and D. Outram, eds., pp. 19–30.

1989. *The Body and the French Revolution: Sex, Class and Political Culture.* New Haven: Yale University Press.

Owen, Alex. 1987. Women and Nineteenth-Century Spiritualism: Strategies in the Subversion of Femininity. In J. Obelkevich, L. Roper, and R. Samuel, eds. *Disciplines of Faith: Studies in Religion, Politics and Patriarchy.* London: Routledge and Kegan Paul, pp. 130–53.

Owen, Richard. 1849. *On the Nature of Limbs.* London: J. Van Voorst.

1855. *Lectures on the Comparative Anatomy and Physiology of the Invertebrate Animals.* 2nd ed. London: Longmans.

1858a. Presidential Address. *Report of the 28th Meeting of the British Association for the Advancement of Science Held at Leeds*, pp. xlix–cx.

1858b. On the Characters, Principles of Division, and Primary Groups of the Class of Mammalia. *Journal of the Proceedings of the Linnean Society* 2: 1–37.

1860. Darwin on the Origin of Species. *Edinburgh Review* 111: 487–532.

1862. *On the Extent and Aims of a National Museum of Natural History.* London: Saunders and Otley.

ed. 1861. *Essays and Observations on Natural History, Anatomy, Physiology, Psychology, and Geology.* By John Hunter. 2 vols. London: J. Van Voorst.

Owen, Richard Starton, ed. 1894. *The Life of Richard Owen.* 2 vols. London: Murray.

Paradis, James. 1978. *T. H. Huxley: Man's Place in Nature.* Lincoln: University of Nebraska Press.

and George Williams. 1989. *Evolution and Ethics: T. H. Huxley's "Evolution and Ethics" with New Essays on Its Victorian and Sociobiological Context.* Princeton: Princeton University Press.

and Thomas Postlewait, eds. 1981. *Victorian Science and Victorian Values: Literary Perspectives.* New York: New York Academy of Sciences.

Perkin, Harold. 1989. *The Rise of Professional Society.* London: Routledge.

Peterson, M. Jeanne. 1989. *Family, Love, and Work in the Lives of Victorian Gentlewomen.* Bloomington: Indiana University Press.

Pick, Daniel. 1989. *Faces of Degeneration.* Cambridge: Cambridge University Press.

Pickstone, John. 1994. Museological Science? The Place of the Analytical/Comparative in Nineteenth-Century Science, Technology, and Medicine. *History of Science* 32: 111–38.

Playfair, Lyon. 1870. *On Primary and Technical Education.* Edinburgh: Edmonston and Douglas.

1888. Lord Armstrong and Technical Education. *Nineteenth Century* 24: 325–33.

Poovey, Mary. 1984. *The Proper Lady and the Woman Writer.* Chicago: University of Chicago Press.

1988. *Uneven Developments: The Ideological Work of Gender in Mid-Victorian England.* Chicago: University of Chicago Press.

Pratt, Mary Louise. 1992. *Imperial Eyes: Travel Writing and Transculturation.* London: Routledge.

Preece, R. C., and I. J. Killeen. 1995. Edward Forbes and Clement Reid: Two Generations of Pioneering Polymaths. *Archives of Natural History* 22: 419–35.

Prochaska, F. K. 1980. *Women and Philanthropy in Nineteenth-Century England.* Oxford: Clarendon Press.

Pycior, Helena, Nancy Slack, and Pnina Abir-Am, eds. 1996. *Creative Couples in the Sciences.* New Brunswick: Rutgers University Press.

Raby, Peter. 2001. *Alfred Russel Wallace: A Life.* London: Chatto and Windus.

Raven, J., H. Small, and N. Tadmor, eds. 1996. *The Practice and Representation of Reading in England.* Cambridge: Cambridge University Press.

Reader, W. J. 1966. *Professional Men: The Rise of the Professional Classes in Nineteenth-Century England.* New York: Basic Books.

Reardon, B. G. 1982. *Liberal Theology in the Nineteenth Century.* Cambridge: Cambridge University Press.

Rehbock, Philip F. 1979. Edward Forbes: An Annotated List of Published and Unpublished Writings. *Journal of the Society for the Bibliography of Natural History* 9: 171–218.

Richards, Evelleen. 1983. Darwin and the Descent of Woman. In D. Oldroyd and I. Langham, eds. *The Wider Domain of Evolutionary Thought.* Dordrecht: D. Reidel, pp. 57–111.

1989. Huxley and Women's Place in Science: The "Woman Question" and the Control of Victorian Anthropology. In J. R. Moore, ed. *History, Humanity and Evolution: Essays for John C. Greene.* Cambridge: Cambridge University Press, pp. 253–84.

1997. Redrawing the Boundaries: Darwinian Science and Victorian Women Intellectuals. In Lightman, ed., pp. 119–42.

Richards, Robert J. 1987. *Darwin and the Emergence of Evolutionary Theories of Mind and Behavior.* Chicago: University of Chicago Press.

1992. *The Meaning of Evolution: The Morphological Construction and Ideological Reconstruction of Darwin's Theory.* Chicago: University of Chicago Press.

Roach, John. 1986. *A History of Secondary Education in England, 1800–1870.* London: Longman.

Roderick, G. W. 1967. *The Emergence of a Scientific Society, 1800–1965.* London: Macmillan.

and M. D. Stephens. 1972. *Scientific and Technical Education in Nineteenth Century England.* Newton Abbot: David and Charles.

Roos, David. 1977. Matthew Arnold and Thomas Henry Huxley: Two Speeches at the Royal Academy, 1881 and 1883. *Modern Philology* 74: 316–24.

1979. "Matthew Arnold, Thomas Henry Huxley, and the Rhetoric of Friendship and Controversy." Ph.D. dissertation, University of Chicago, Department of English.

Roper, Michael, and John Tosh, eds. 1991. *Manful Assertions: Masculinities in Britain since 1800.* London: Routledge.

Ross, Sidney. 1962. "Scientist": The Story of a Word. *Annals of Science* 18: 65–85.

Rothblatt, Sheldon. 1968. *The Revolution of the Dons: Cambridge and Society in Victorian England*. New York: Basic Books.

1976. *Tradition and Change in English Liberal Education*. London: Faber and Faber.

Rudwick, Martin J. S. 1982. Charles Darwin in London: The Integration of Public and Private Science. *Isis* 73: 186–206.

1985. *The Great Devonian Controversy*. Chicago: University of Chicago Press.

Rupke, Nicolaas. 1983. *The Great Chain of History: William Buckland and the English School of Geology (1814–1849)*. Oxford: Clarendon.

1994. *Richard Owen: Victorian Naturalist*. New Haven: Yale University Press.

Ruse, Michael. 1979. *The Darwinian Revolution: Science Red in Tooth and Claw*. Chicago: University of Chicago Press.

Russell-Gebbett, Jean. 1977. *Henslow of Hitcham: Botanist, Educationalist and Clergyman*. Lavenham, Suffolk: Terence Dalton.

Russett, Cynthia. 1989. *Sexual Science: The Victorian Construction of Womanhood*. Cambridge: Harvard University Press.

St. George, E. A. W. 1993. *The Descent of Manners: Etiquette, Rules and the Victorians*. London: Chatto and Windus.

Sandall, Robert. 1955. *The History of the Salvation Army*. 3 vols. London: Thomas Nelson and Sons.

Sanders, Charles. 1942. *Coleridge and the Broad Church Movement*. Durham: Duke University Press.

Sanford, Frances R., ed. 1889. *Reports on Elementary Schools, 1852–1882*. By Matthew Arnold. London: Macmillan.

Schaffer, Simon. 1988. Astronomers Mark Time: Discipline and the Personal Equation. *Science in Context* 2: 115–45.

1990. Genius in Romantic Natural Philosophy. In A. Cunningham and N. Jardine, eds. *Romanticism and the Sciences*. Cambridge: Cambridge University Press, pp. 82–98.

1992. Late Victorian Metrology and Its Instrumentation: A Manufactory of Ohms. In R. Bud and S. Cozzens, eds. *Invisible Connections: Instruments, Institutions, and Science*. Bellingham, WA: SPIE Optical Engineering Press, pp. 23–56.

1995. Accurate Measurement Is an English Science. In M. Norton Wise, ed. *The Values of Precision*. Princeton: Princeton University Press, pp. 135–72.

Schiebinger, Londa. 1989. *The Mind Has No Sex? Women in the Origin of Modern Science*. Cambridge: Harvard University Press.

Scott, Joan. 1983. Women in History: The Modern Period. *Past and Present* 101: 141–57.

Searle, G. R. 1976. *Eugenics and Politics in Britain*. Leyden: Noordhoff International.

Secord, Anne. 1994a. Science in the Pub: Artisan Botany in Early Nineteenth-Century Lancashire. *History of Science* 32: 269–315.

1994b. Corresponding Interests: Artisans and Gentlemen in Nineteenth-Century Natural History. *British Journal for the History of Science* 27: 383–408.

2002. Botany on a Plate: Pleasure and the Power of Pictures in Promoting Early Nineteenth-Century Scientific Knowledge. *Isis*: in press.

Secord, James A. 1982. King of Siluria: Roderick Murchison and the British Imperial Theme in Nineteenth-Century British Geology. *Victorian Studies* 25: 413–42.

1985. John W. Salter: The Rise and Fall of a Victorian Palaeontological Career. In A. Wheeler and J. Price, eds. *From Linnaeus to Darwin: Commentaries on the History of Biology and Geology.* London: Society for the History of Natural History, pp. 61–75.

1986a. *Controversy in Victorian Geology.* Princeton: Princeton University Press.

1986b. The Geological Survey of Great Britain as a Research School, 1839–1855. *History of Science* 24: 223–75.

2000. *Victorian Sensation: The Extraordinary Reception, Publication, and Secret Authorship of* Vestiges of the Natural History of Creation. Chicago: University of Chicago Press.

Sedwick, Adam. 1850. *A Discourse on the Studies of the University of Cambridge.* 5th ed. London: John W. Parker.

Shaffer, E. S., ed. 1998. *The Third Culture: Literature and Science.* New York: Walter de Gruyter.

Shapin, Steven. 1988. The House of Experiment in 17th Century England. *Isis* 79: 373–404.

1990. "The Mind Is Its Own Place": Science and Solitude in Seventeenth-Century England. *Science in Context* 4: 191–218.

1991. "A Scholar and a Gentleman": The Problematic Identity of the Scientific Practitioner in Early Modern England. *History of Science* 29: 279–327.

1994. *A Social History of Truth: Civility and Science in Seventeenth-Century England.* Chicago: University of Chicago Press.

and Barry Barnes. 1977. Science, Nature and Control: Interpreting Mechanics' Institutes. *Social Studies of Science* 7: 31–74.

Shattock, J., and M. Wolff, eds. 1982. *The Victorian Periodical Press: Samplings and Surroundings.* Leicester: Leicester University Press.

Shea, Victor, and William Whitla, eds. 2000. Essays and Reviews: *The 1860 Text and Its Reading.* Charlottesville: University Press of Virginia.

Sheets-Pyenson, Susan. 1985. Popular Scientific Periodicals in Paris and London: The Emergence of a Low Scientific Culture, 1820–1875. *Annals of Science* 42: 549–72.

1988. *Cathedrals of Science: The Development of Colonial Natural History Museums during the Nineteenth Century.* Kingston, Ont.: McGill-Queen's University Press.

Shrosbree, Colin. 1988. *Public Schools and Private Education: The Clarendon Commission, 1861–1864, and the Public Schools Acts.* Manchester: Manchester University Press.

Shuttleworth, Sally. 1984. *George Eliot and Nineteenth Century Science.* Cambridge: Cambridge University Press.

1990. Female Circulation: Medical Discourse and Popular Advertising in the Mid-Victorian Era. In M. Jacobus, E. Keller, and S. Shuttleworth, eds.

Body/Politics: Women and the Discourses of Science. London: Routledge, pp. 47–63.

1996. *Charlotte Bronte and Victorian Psychology.* Cambridge: Cambridge University Press.

Simon, Brian. 1960. *Studies in the History of Education, 1780–1870.* London: Lawrence and Wishart.

1965. *Education and the Labour Movement, 1870–1920.* London: Lawrence and Wishart.

1987. Systematization and Segmentation in Education: The Case of England. In D. Müller, F. Ringer, and B. Simon, eds., pp, 88–108.

Small, Helen. 1994. "In the Guise of Science": Literature and Rhetoric of Nineteenth-Century English Psychiatry. *History of the Human Sciences* 7: 27–55.

Smiles, Samuel. 1871. *Character.* London: Murray.

Smith, Bernard. 1985. *European Vision and the South Pacific.* New Haven: Yale University Press.

Smith, Crosbie, and M. Norton Wise. 1989. *Energy and Empire: A Biographical Study of Lord Kelvin.* Cambridge: Cambridge University Press.

Snow, C. P. 1993. *The Two Cultures.* Cambridge: Cambridge University Press.

Spencer, Herbert. 1851. *Social Statics.* London: Chapman.

1852a. Use and Beauty. *The Leader,* 3 January.

1852b. The Philosophy of Style. *Westminster Review* 58: 435–59.

1860–2. *First Principles.* London: Williams and Norgate.

1861. What Knowledge Is of Most Worth? In *Education: Intellectual, Moral and Physical.* London: Williams and Norgate, pp. 1–55.

1870–2. *The Principles of Psychology.* 2nd ed. 2 vols. London: Williams and Norgate.

1876–96. *The Principles of Sociology.* 3 vols. London: Williams and Norgate.

Stafford, R. A. 1989. *Scientist of Empire: Sir Roderick Murchison, Scientific Exploration and Victorian Imperialism.* Cambridge: Cambridge University Press.

Stanley, Arthur Penrhyn. 1863. *Letter to the Bishop of London on the State of Subscription in the Church of England, and in the University of Oxford.* Oxford: John Henry.

Steedman, Hilary. 1987. Defining Institutions: The Endowed Grammar Schools and the Systematization of English Secondary Education. In D. Muller, F. Ringer, and B. Simon, eds., pp. 111–34.

Stephen, Leslie, and Sidney Lee, eds. 1885–1912. *Dictionary of National Biography.* 63 vols. London: Smith, Elder, and Co.

Stevens, Peter, ed. 1993. *City and Guilds of London Institute: A Short History, 1878–1992.* London: City and Guilds of London Institute.

Stocking, George, ed. 1985. *Objects and Others: Essays on Museums and Material Culture.* Madison: University of Wisconsin Press.

Strick, James. 2000. *Sparks of Life: Darwinism and the Victorian Debates over Spontaneous Generation.* Cambridge, MA: Harvard University Press.

Super, R. H. 1977. The Humanist at Bay: The Arnold-Huxley Debate. In U. C. Knoepflmacher and G. B. Tennyson, eds. *Nature and the Victorian Imagination.* Berkeley: University of California Press, pp. 231–45.

ed. 1960–77. *Complete Prose Works of Matthew Arnold*. 12 vols. Ann Arbor: University of Michigan Press.

Taylor, M. W. 1992. *Man Versus the State: Herbert Spencer and Late Victorian Liberalism*. Oxford: Clarendon.

Temple, William. 1884. *Bampton Lectures on Science and Religion*. Cambridge: Cambridge University Press.

Thackray, Arnold. 1974. The Industrial Revolution and the Image of Science. In A. Thackray and E. Mendelsohn, eds. *Science and Values*. New York: Humanities Press, pp. 3–18.

Thompson, E. P. 1963. *The Making of the English Working Class*. London: Pantheon.

Thompson, F. M. L. 1988. *The Rise of Respectable Society*. Cambridge, MA: Harvard University Press.

Tosh, John. 1991. Domesticity and Manliness in the Victorian Middle-Class. In M. Roper and J. Tosh, eds., pp. 44–73.

Turner, Frank. 1974. *Between Science and Religion: The Reaction to Scientific Naturalism in Late Victorian England*. New Haven: Yale University Press.

1981. *The Greek Heritage in Victorian Britain*. New Haven: Yale University Press.

1993. *Contesting Cultural Authority: Essays in Victorian Intellectual Life*. Cambridge: Cambridge University Press.

Tyndall, John. 1868. *Faraday as a Discoverer*. London: Longmans, Green, and Co.

1870. On the Scientific Use of the Imagination. In *Fragments of Science*, vol. 1. London: Longmans, Green, and Co., 1871, pp. 126–67.

1874. The Belfast Address. In *Fragments of Science*, vol. 2. London: Longmans, Green, and Co., 1879, pp. 137–203.

Vance, Norman. 1985. *The Sinews of the Spirit: The Ideal of Christian Manliness in Victorian Literature and Religious Thought*. Cambridge: Cambridge University Press.

Vicinus, Martha, ed. 1972. *Suffer and Be Still: Women in the Victorian Age*. Bloomington: Indiana University Press.

Vickery, Amanda. 1993. Golden Age to Separate Spheres? A Review of the Categories and Chronology of English Women's History. *Historical Journal* 36: 383–414.

Vincent-Buffault, Anne. 1991. *The History of Tears: Sensibility and Sentimentality in France*. London: Macmillan.

Wahrman, Dror. 1993. "Middle-Class" Domesticity Goes Public: Gender, Class, and Politics from Queen Caroline to Queen Victoria. *Journal of British Studies* 32: 396–432.

1995. *Imagining the Middle Class: The Political Representation of Class in Britain, 1780–1840*. Cambridge: Cambridge University Press.

Walker, Pamela J. 1991. "I Live But Not Yet I For Christ Liveth In Me": Men and Masculinity in the Salvation Army, 1865–90. In M. Roper and J. Tosh, eds., pp. 92–112.

Ward, Humphrey. 1926. *History of the Athenaeum, 1824–1925*. London: William Clower.

Wee, C. J. W-L. 1994. Christian Manliness and National Identity: The Problematic Construction of the Racially "Pure" Nation. In D. E. Hall, ed. *Muscular Christianity: Embodying the Victorian Age*. Cambridge: Cambridge University Press, pp. 66–88.

Wheeler-Barclay, Margorie. 1987. "The Science of Religion in Britain, 1860–1915." Ph.D. dissertation, Northwestern University, Department of History.

White, Paul. 2000. Thomas Huxley. In Arne Hessenbruch, ed. *Reader's Guide to the History of Science*. London: Fitzroy Dearborn.

Williams, Raymond. 1958. *Culture and Society*. London: Chatto and Windus.

Williams, Rowland. 1861–2. *Hints to My Counsel in the Court of Arches*. London: Taylor and Francis.

Winsor, Mary P. 1976. *Starfish, Jellyfish, and the Order of Life*. New Haven: Yale University Press.

Winter, Alison. 1998a. *Mesmerized: Powers of Mind in Victorian Britain*. Chicago: University of Chicago Press.

1998b. A Calculus of Suffering: Ada Lovelace and the Bodily Constraints on Women's Knowledge in Early Victorian England. In C. Lawrence and S. Shapin, eds. *Science Incarnate: Historical Embodiments of Natural Knowledge*. Chicago: University of Chicago Press, pp. 202–39.

Wright, Terence. 1986. *The Religion of Humanity: The Impact of Comtean Positivism on Victorian Britain*. Cambridge: Cambridge University Press.

Yanni, Carla. 1999. *Nature's Museums: Victorian Science and the Architecture of Display*. London: Athlone.

Yeo, Eileen. 1971. Mayhew as a Social Investigator. In E. P. Thompson and E. Yeo, eds. *The Unknown Mayhew: Selections from the Morning Chronicle, 1849–1850*. London: Merlin Press.

1987. Chartist Religious Belief and the Theology of Liberation. In J. Obelkevich, L. Roper, and R. Samuel, eds. *Disciplines of Faith: Studies in Religion, Politics and Patriarchy*. London: Routledge and Kegan Paul, pp. 410–21.

Yeo, Richard. 1993. *Defining Science: William Whewell, Natural Knowledge, and Public Debate in Early Victorian Britain*. Cambridge: Cambridge University Press.

Young, Robert M. 1985. *Darwin's Metaphor: Nature's Place in Victorian Culture*. Cambridge: Cambridge University Press.

Index